I0503201

The Flavour of Particle Physics

Edited by

Paul F. Kisak

Virginia, USA

Visit our website at: https://www.createspace.com/5814842

Printed in The United States of America

First Trade Edition: 2015
10 9 8 7 6 5 4 3 2 1

Black & White on White paper
65 pages

ISBN-13: 978-1518713408
ISBN-10: 1518713408

Virginia, USA

Contents

Chapter 1

Flavour (particle physics)

In particle physics, **flavour** or **flavor** refers to a species of an elementary particle. The Standard Model counts six flavours of quarks and six flavours of leptons. They are conventionally parameterized with *flavour quantum numbers* that are assigned to all subatomic particles, including composite ones. For hadrons, these quantum numbers depend on the numbers of constituent quarks of each particular flavour.

1.1 Intuitive description

Elementary particles are not eternal and indestructible. Unlike in classical mechanics, where forces only change a particle's momentum, the weak force can alter the essence of a particle, even an elementary particle. This means that it can convert one quark to another quark with different mass and electric charge, and the same for leptons. From the point of view of quantum mechanics, changing the flavour of a particle by the weak force is no different in principle from changing its spin by electromagnetic interaction, and should be described with quantum numbers as well. In particular, flavour states may undergo quantum superposition.

In atomic physics the principal quantum number of an electron specifies the electron shell in which it resides, which determines the energy level of the whole atom. In an analogous way, the five flavour quantum numbers of a quark specify which of six flavours (u, d, s, c, b, t) it has, and when these quarks are combined this results in different types of baryons and mesons with different masses, electric charges, and decay modes.

1.2 Flavour symmetry

If there are two or more particles which have identical interactions, then they may be interchanged without affecting the physics. Any (complex) linear combination of these two particles give the same physics, as long as they are orthogonal or perpendicular to each other. In other words, the theory possesses symmetry transformations such as $M \left(\begin{smallmatrix} u \\ d \end{smallmatrix} \right)$, where u and d are the two fields, and M is any 2×2 unitary matrix with a unit determinant. Such matrices form a Lie group called SU(2) (see special unitary group). This is an example of flavour symmetry.

In quantum chromodynamics, flavour is a global symmetry. In the electroweak theory, on the other hand, this symmetry is broken, and flavour changing processes exist, such as quark decay or neutrino oscillations.

1.3 Flavour quantum numbers

1.3.1 Leptons

All leptons carry a lepton number $L = 1$. In addition, leptons carry weak isospin, T_3, which is $-1/2$ for the three charged leptons (i.e. electron, muon and tau) and $+1/2$ for the three associated neutrinos. Each doublet of a charged lepton and a neutrino consisting of opposite T_3 are said to constitute one generation of leptons. In addition, one defines a quantum number called weak hypercharge, YW, which is -1 for all left-handed leptons.[1] Weak isospin and weak hypercharge are gauged in the Standard Model.

Leptons may be assigned the six flavour quantum numbers: electron number, muon number, tau number, and corresponding numbers for the neutrinos. These are conserved in strong and electromagnetic interactions, but violated by weak interactions. Therefore, such flavour quantum numbers are not of great use. A separate quantum number for each generation is more useful: electronic lepton number (+1 for electrons and electron neutrinos), muonic lepton number (+1 for muons and muon neutrinos), and tauonic lepton number (+1 for tau leptons and tau neutrinos). However, even these numbers are not absolutely conserved, as neutrinos of different generations can mix; that is, a neutrino of one flavour can transform into another flavour. The strength of such mixings is specified by a matrix called the Pontecorvo–Maki–Nakagawa–Sakata matrix (PMNS matrix).

1.3.2 Quarks

All quarks carry a baryon number $B = 1/3$. They also all carry weak isospin, $T_3 = \pm1/2$. The positive-T_3 quarks (up, charm, and top quarks) are called *up-type quarks* and negative-T_3 quarks (down, strange, and bottom quarks) are called *down-type quarks*. Each doublet of up and down type quarks constitutes one generation of quarks.

For all the quark flavour quantum numbers (strangeness, charm, topness and bottomness) the convention is that the flavour charge and the electric charge of a quark have the same sign. Thus any flavour carried by a charged meson has the same sign as its charge. Quarks have the following flavour quantum numbers:

- Isospin, less ambiguously known as "isobaric spin", which has value $I_3 = 1/2$ for the up quark and $I_3 = -1/2$ for the down quark.

- Strangeness (S): Defined as $S = -(n_s - \bar{n_s})$, where n_s represents the number of strange quarks (s) and $\bar{n_s}$ represents the number of strange antiquarks (s). This quantum number was introduced by Murray Gell-Mann. This definition gives the strange quark a strangeness of -1 for the above-mentioned reason.

- Charm (C): Defined as $C = (n_c - \bar{n_c})$, where n_c represents the number of charm quarks (c) and $\bar{n_c}$ represents the number of charm antiquarks. Is +1 for the charm quark.

- Bottomness (B'): Also called 'beauty'. Defined as $B' = -(n_b - \bar{n_b})$, where n_b represents the number of bottom quarks (b) and $\bar{n_b}$ represents the number of bottom antiquarks.

- Topness (T): Also called 'truth'. Defined as $T = (n_t - \bar{n_t})$, where n_t represents the number of top quarks (t) and $\bar{n_t}$ represents the number of top antiquarks. However, because of the extremely short half-life of the top quark, by the time it can interact strongly it has already decayed to another flavour of quark (usually to a bottom quark). For that reason the top quark doesn't hadronize, that is it never forms any meson or baryon.

These five quantum numbers, together with baryon number (which is not a flavour quantum number) completely specify numbers of all 6 quark flavours separately (as $n_q - \bar{n_q}$, i.e. an antiquark is counted with the minus sign). They are conserved by both the electromagnetic and strong interactions (but not the weak interaction). From them can be built the derived quantum numbers:

- Hypercharge (Y): $Y = B + S + C + B' + T$

- Electric charge: $Q = I_3 + 1/2Y$ (see Gell-Mann–Nishijima formula)

The terms "strange" and "strangeness" predate the discovery of the quark, but continued to be used after its discovery for the sake of continuity (i.e. the strangeness of each type of hadron remained the same); strangeness of anti-particles being referred to as +1, and particles as −1 as per the original definition. Strangeness was introduced to explain the rate of decay of newly discovered particles, such as the kaon, and was used in the Eightfold Way classification of hadrons and in subsequent quark models. These quantum numbers are preserved under strong and electromagnetic interactions, but not under weak interactions.

For first-order weak decays, that is processes involving only one quark decay, these quantum numbers (e.g. charm) can only vary by 1 ($|C| = \pm1$); $\Delta B' = \pm1$. Since first-order processes are more common than second-order processes (involving two quark decays), this can be used as an approximate "selection rule" for weak decays.

A quark of a given flavour is an eigenstate of the weak interaction part of the Hamiltonian: it will interact in a definite way with the W and Z bosons. On the other hand, a fermion of a fixed mass (an eigenstate of the kinetic and strong interaction parts of the Hamiltonian) is normally a superposition of various flavours. As a result, the flavour content of a quantum state may change as it propagates freely. The transformation from flavour to mass basis for quarks is given by the Cabibbo–Kobayashi–Maskawa matrix (CKM matrix). This matrix is analogous to the PMNS matrix for neutrinos, and defines the strength of flavour changes under weak interactions of quarks.

The CKM matrix allows for CP violation if there are at least three generations.

1.3.3 Antiparticles and hadrons

Flavour quantum numbers are additive. Hence antiparticles have flavour equal in magnitude to the particle but opposite in sign. Hadrons inherit their flavour quantum number from their valence quarks: this is the basis of the classification in the quark model. The relations between the hypercharge, electric charge and other flavour quantum numbers hold for hadrons as well as quarks.

1.4 Quantum chromodynamics

Flavour symmetry is closely related to chiral symmetry. This part of the article is best read along with the one on chirality.

Quantum chromodynamics (QCD) contains six flavours of quarks. However, their masses differ and as a result they are not strictly interchangeable with each other. The up and down flavours are close to having equal masses, and the theory of these two quarks possesses an approximate SU(2) symmetry (isospin symmetry).

Under some circumstances, the masses of the quarks can be neglected entirely. One can then make flavour transformations independently on the left- and right-handed parts of each quark field. The flavour group is then a chiral group SUL(N_f) × SUR(N_f).

If all quarks had non-zero but equal masses, then this chiral symmetry is broken to the *vector symmetry* of the "diagonal flavour group" SU(N_f), which applies the same transformation to both helicities of the quarks. Such a reduction of the symmetry is called *explicit symmetry breaking*. The amount of explicit symmetry breaking is controlled by the current quark masses in QCD.

Even if quarks are massless, chiral flavour symmetry can be spontaneously broken if the vacuum of the theory contains a chiral condensate (as it does in low-energy QCD). This gives rise to an effective mass for the quarks, often identified with the valence quark mass in QCD.

1.4.1 Symmetries of QCD

Analysis of experiments indicate that the current quark masses of the lighter flavours of quarks are much smaller than the QCD scale, ΛQCD, hence chiral flavour symmetry is a good approximation to QCD for the up, down and strange quarks. The success of chiral perturbation theory and the even more naive chiral models spring from this fact. The valence quark masses extracted from the quark model are much larger than the current quark mass. This indicates that QCD has spontaneous chiral symmetry breaking with the formation of a chiral condensate. Other phases of QCD may break the chiral flavour symmetries in other ways.

1.5   Conservation laws

All of the various charges discussed above are conserved by the fact that the charge operator is best understood as the generator of a symmetry that commutes with the Hamiltonian. Thus, the eigenvalues of the various charge operators are conserved.

Absolutely conserved flavour quantum numbers are: (including the baryon number for completeness)

- electric charge (Q)
- weak isospin (I_3)
- baryon number (B)
- lepton number (L)

In some theories, the individual baryon and lepton number conservation can be violated, if the difference between them ($B - L$) is conserved (see chiral anomaly). All other flavour quantum numbers are violated by the electroweak interactions. Strong interactions conserve all flavours.

1.6   History

Some of the historical events that lead to the development of flavour symmetry are discussed in the article on isospin.

1.7   See also

- Standard Model (mathematical formulation)
- Cabibbo–Kobayashi–Maskawa matrix
- Strong CP problem and chirality (physics)
- Chiral symmetry breaking and quark matter
- Quark flavour tagging, such as B-tagging, is an example of particle identification in experimental particle physics.

1.8   References

[1] See table in S. Raby, R. Slanky (1997). "Neutrino Masses: How to add them to the Standard Model" (PDF). *Los Alamos Science* (25): 64.

1.9   Further reading

- Lessons in Particle Physics Luis Anchordoqui and Francis Halzen, University of Wisconsin, 18th Dec. 2009

1.10   External links

- The particle data group.

Chapter 2

Quantum number

"Q-number" redirects here. For the q-theory concept, see q-analog.

Quantum numbers describe values of conserved quantities in the dynamics of a quantum system. In the case of quantum numbers of electrons, they can be defined as "The sets of numerical values which give acceptable solutions to the Schrödinger wave equation for the hydrogen atom". Perhaps the most important aspect of quantum mechanics is the quantization of observable quantities, since quantum numbers are discrete sets of integers or half-integers, although they could approach infinity in some cases. This is distinguished from classical mechanics where the values can range continuously. Quantum numbers often describe specifically the energy levels of electrons in atoms, but other possibilities include angular momentum, spin, etc. Any quantum system can have one or more quantum numbers; it is thus difficult to list all possible quantum numbers.[1]

2.1 How many quantum numbers?

The question of *how many quantum numbers are needed to describe any given system* has no universal answer. Hence for each system one must find the answer for a full analysis of the system. A quantized system requires at least one quantum number. The dynamics of any quantum system are described by a quantum Hamiltonian, H. There is one quantum number of the system corresponding to the energy, i.e., the eigenvalue of the Hamiltonian. There is also one quantum number for each operator O that commutes with the Hamiltonian. These are all the quantum numbers that the system can have. Note that the operators O defining the quantum numbers should be independent of each other. Often, there is more than one way to choose a set of independent operators. Consequently, in different situations different sets of quantum numbers may be used for the description of the same system.

2.2 Spatial and angular momentum numbers

There are four quantum numbers which can describe an electron in an atom completely.

- Principal quantum number (n)

- Azimuthal quantum number (ℓ)

- Magnetic quantum number (m)

- Spin quantum number (s)

2.2.1 Traditional nomenclatures

Many different models have been proposed throughout the history of quantum mechanics, but the most prominent system of nomenclature spawned from the Hund-Mulliken molecular orbital theory of Friedrich Hund, Robert S. Mulliken, and contributions from Schrödinger, Slater and John Lennard-Jones. This system of nomenclature incorporated Bohr energy levels, Hund-Mulliken orbital theory, and observations on electron spin based on spectroscopy and Hund's rules.[2]

This model describes electrons using four quantum numbers, n, ℓ, $m\ell$, ms, given below. It is also the common nomenclature in the classical description of nuclear particle states (e.g. protons and neutrons). Molecular orbitals require different quantum numbers, because the Hamiltonian and its symmetries are quite different.

1. **The principal quantum number (n)** describes the electron shell, or energy level, of an atom. The value of n ranges from 1 to the shell containing the outermost electron of that atom, i.e.[3]

   $$n = 1, 2, \dots \, .$$

   For example, in caesium (Cs), the outermost valence electron is in the shell with energy level 6, so an electron in caesium can have an n value from 1 to 6.

   For particles in a time-independent potential (see Schrödinger equation), it also labels the nth eigenvalue of Hamiltonian (H), i.e. the energy, E with the contribution due to angular momentum (the term involving $\mathbf{J}^2$) left out. This number therefore has a dependence only on the distance between the electron and the nucleus (i.e., the radial coordinate, $\mathbf{r}$). The average distance increases with $\mathbf{n}$, and hence quantum states with different principal quantum numbers are said to belong to different shells.

2. **The azimuthal quantum number (ℓ)** (also known as the **angular quantum number** or **orbital quantum number**) describes the subshell, and gives the magnitude of the orbital angular momentum through the relation

   $$L^2 = \hbar^2 \, \ell \, (\ell + 1).$$

   In chemistry and spectroscopy, "$\ell = 0$" is called an s orbital, "$\ell = 1$" a p orbital, "$\ell = 2$" a d orbital, and "$\ell = 3$" an f orbital.

   The value of ℓ ranges from 0 to $n - 1$, because the first p orbital ($\ell = 1$) appears in the second electron shell ($n = 2$), the first d orbital ($\ell = 2$) appears in the third shell ($n = 3$), and so on:[4]

   $$\ell = 0, 1, 2, \dots, n - 1.$$

   A quantum number beginning in 3, 0, ... describes an electron in the s orbital of the third electron shell of an atom. In chemistry, this quantum number is very important, since it specifies the shape of an atomic orbital and strongly influences chemical bonds and bond angles.

3. **The magnetic quantum number ($m\ell$)** describes the specific orbital (or "cloud") within that subshell, and yields the *projection* of the orbital angular momentum *along a specified axis*:

   $$Lz = m\ell \, \hbar.$$

   The values of $m\ell$ range from $-\ell$ to ℓ, with integer steps between them:[5]

   The s subshell ($\ell = 0$) contains only one orbital, and therefore the $m\ell$ of an electron in an s orbital will always be 0. The p subshell ($\ell = 1$) contains three orbitals (in some systems, depicted as three "dumbbell-shaped" clouds), so the $m\ell$ of an electron in a p orbital will be -1, 0, or 1. The d subshell ($\ell = 2$) contains five orbitals, with $m\ell$ values of -2, -1, 0, 1, and 2.

4. **The spin projection quantum number (ms)** describes the spin (intrinsic angular momentum) of the electron within that orbital, and gives the projection of the spin angular momentum S along the specified axis:

   $$Sz = ms \, \hbar.$$

   In general, the values of ms range from $-s$ to s, where s is the spin quantum number, an intrinsic property of particles:[6]

$ms = -s, -s + 1, -s + 2,...,s - 2, s - 1, s.$

An electron has spin number $s = \frac{1}{2}$, consequently ms will be $\pm\frac{1}{2}$, referring to "spin up" and "spin down" states. Each electron in any individual orbital must have different quantum numbers because of the Pauli exclusion principle, therefore an orbital never contains more than two electrons.

Note that there is no universal fixed value for $m\ell$ and ms values. Rather, the $m\ell$ and ms values are random. The only requirement is that the naming schematic used within a particular set of calculations or descriptions must be consistent (e.g. the orbital occupied by the first electron in a p orbital could be described as $m\ell = -1$ or $m\ell = 0$, or $m\ell = 1$, but the $m\ell$ value of the other electron in that orbital must be different; yet, the $m\ell$ assigned to electrons in other orbitals again can be $m\ell = -1$ or $m\ell = 0$, or $m\ell = 1$).

These rules are summarized as follows:

Example: The quantum numbers used to refer to the outermost valence electrons of the Carbon (C) atom, which are located in the 2p atomic orbital, are; $n = 2$ (2nd electron shell), $\ell = 1$ (p orbital subshell), $m\ell = 1, 0$ or -1, $ms = \frac{1}{2}$ (parallel spins).

Results from spectroscopy indicated that up to two electrons can occupy a single orbital. However two electrons can never have the same exact quantum state nor the same set of quantum numbers according to Hund's rules, which addresses the Pauli exclusion principle. A fourth quantum number with two possible values was added as an *ad hoc* assumption to resolve the conflict; this supposition could later be explained in detail by relativistic quantum mechanics and from the results of the renowned Stern–Gerlach experiment.

2.3 Total angular momenta numbers

2.3.1 Total momentum of a particle

For more details on this topic, see Clebsch–Gordan coefficients.
See also: Azimuthal quantum number § Total angular momentum of an electron in the atom

When one takes the spin-orbit interaction into consideration, the L and S operators no longer commute with the Hamiltonian, and their eigenvalues therefore change over time. Thus another set of quantum numbers should be used. This set includes[7][8]

1. **The total angular momentum quantum number:**

   $j = |\ell \pm s|$

   which gives the total angular momentum through the relation

   $J^2 = \hbar^2 j (j + 1).$

2. **The projection of the total angular momentum along a specified axis:**

   $mj = -j, -j + 1, -j + 2,...,j - 2, j - 1, j$

   analogous to the above, and satisfies

   $mj = m\ell + ms$ and $|m\ell + ms| \leq j.$

3. **Parity**

This is the eigenvalue under reflection, and is positive (+1) for states which came from even ℓ and negative (−1) for states which came from odd ℓ. The former is also known as **even parity** and the latter as **odd parity**, and is given by

$$P = (-1)^{\ell}.$$

For example, consider the following eight states, defined by their quantum numbers:

The quantum states in the system can be described as linear combination of these eight states. However, in the presence of spin-orbit interaction, if one wants to describe the same system by eight states which are eigenvectors of the Hamiltonian (i.e. each represents a state which does not mix with others over time), we should consider the following eight states:

2.3.2 Nuclear angular momentum quantum numbers

In nuclei, the entire assembly of protons and neutrons (nucleons) has a resultant angular momentum due to the angular momenta of each nucleon, usually denoted **I**. If the total angular momentum of a neutron is $jn = \ell + s$ and for a proton is $jp = \ell + s$ (where s for protons and neutrons happens to be ½ again) then the **nuclear angular momentum quantum numbers** I are given by:

$$I = |jn - jp|, |jn - jp| + 1, |jn - jp| + 2,..., (jn + jp) - 2, (jn + jp) - 1, (jn + jp)$$

Parity with the number I is used to label nuclear angular momentum states, examples for some isotopes of Hydrogen (H), Carbon (C), and Sodium (Na) are;[9]

The reason for the unusual fluctuations in I, even by differences of just one nucleon, are due to the odd/even numbers of protons and neutrons - pairs of nucleons have a total angular momentum of zero (just like electrons in orbitals), leaving an odd/even numbers of unpaired nucleons. The property of nuclear spin is an important factor for the operation of NMR spectroscopy in organic chemistry,[8] and MRI in nuclear medicine,[9] due to the nuclear magnetic moment interacting with an external magnetic field.

2.4 Elementary particles

For a more complete description of the quantum states of elementary particles, see Standard model and Flavour (particle physics).

Elementary particles contain many quantum numbers which are usually said to be intrinsic to them. However, it should be understood that the elementary particles are quantum states of the standard model of particle physics, and hence the quantum numbers of these particles bear the same relation to the Hamiltonian of this model as the quantum numbers of the Bohr atom does to its Hamiltonian. In other words, each quantum number denotes a symmetry of the problem. It is more useful in quantum field theory to distinguish between spacetime and internal symmetries.

Typical quantum numbers related to spacetime symmetries are spin (related to rotational symmetry), the parity, C-parity and T-parity (related to the Poincaré symmetry of spacetime). Typical **internal symmetries** are lepton number and baryon number or the electric charge. (For a full list of quantum numbers of this kind see the article on flavour.)

A minor but often confusing point is as follows: most conserved quantum numbers are additive, so in an elementary particle reaction, the *sum* of the quantum numbers should be the same before and after the reaction. However, some, usually called a *parity*, are multiplicative; i.e., their *product* is conserved. All multiplicative quantum numbers belong to a symmetry (like parity) in which applying the symmetry transformation twice is equivalent to doing nothing (involution). These are all examples of an abstract group called $\mathbf{Z}_2$.

2.5 See also

- Electron configuration

2.6 References and external links

[1] McGraw Hill Encyclopaedia of Physics (2nd Edition), C.B. Parker, 1994, ISBN 0-07-051400-3

[2] Chemistry, Matter, and the Universe, R.E. Dickerson, I. Geis, W.A. Benjamin Inc. (USA), 1976, ISBN 0-19-855148-7

[3] Concepts of Modern Physics (4th Edition), A. Beiser, Physics, McGraw-Hill (International), 1987, ISBN 0-07-100144-1

[4] Molecular Quantum Mechanics Parts I and II: An Introduction to QUANTUM CHEMISRTY (Volume 1), P.W. Atkins, Oxford University Press, 1977, ISBN 0-19-855129-0

[5] Quantum Physics of Atoms, Molecules, Solids, Nuclei, and Particles (2nd Edition), R. Eisberg, R. Resnick, John Wiley & Sons, 1985, ISBN 978-0-471-87373-0

[6] Quantum Mechanics (2nd edition), Y. Peleg, R. Pnini, E. Zaarur, E. Hecht, Schuam's Outlines, McGraw Hill (USA), 2010, ISBN 978-0-07-162358-2

[7] Molecular Quantum Mechanics Parts I and II: An Introduction to QUANTUM CHEMISTRY (Volume 1), P.W. Atkins, Oxford University Press, 1977, ISBN 0-19-855129-0

[8] Molecular Quantum Mechanics Part III: An Introduction to QUANTUM CHEMISTRY (Volume 2), P.W. Atkins, Oxford University Press, 1977

[9] Introductory Nuclear Physics, K.S. Krane, 1988, John Wiley & Sons Inc, ISBN 978-0-471-80553-3

2.6.1 General principles

- Dirac, Paul A.M. (1982). *Principles of quantum mechanics.* Oxford University Press. ISBN 0-19-852011-5.

2.6.2 Atomic physics

- Quantum numbers for the hydrogen atom
- Lecture notes on quantum numbers

2.6.3 Particle physics

- Griffiths, David J. (2004). *Introduction to Quantum Mechanics (2nd ed.).* Prentice Hall. ISBN 0-13-805326-X.
- Halzen, Francis and Martin, Alan D. (1984). *QUARKS AND LEPTONS: An Introductory Course in Modern Particle Physics.* John Wiley & Sons. ISBN 0-471-88741-2.
- The particle data group

Chapter 3

Isospin

In nuclear physics and particle physics, **isospin** (*isotopic spin*, *isobaric spin*) is a quantum number related to the strong interaction. Particles that are affected equally by the strong force but have different charges (e.g. protons and neutrons) can be treated as being different states of the same particle with isospin values related to the number of charge states.[1]

Although it does not have the units of angular momentum and is not a type of spin, the formalism that describes it is mathematically similar to that of angular momentum in quantum mechanics, which means it can be coupled in the same manner. For example, a proton-neutron pair can be coupled in a state of total isospin 1 or 0.[2] It is a dimensionless quantity and the name derives from the fact that the mathematical structures used to describe it are very similar to those used to describe the intrinsic angular momentum (spin).

This term was derived from *isotopic spin*, a confusing term to which nuclear physicists prefer *isobaric spin*, which is more precise in meaning. Isospin symmetry is a subset of the flavour symmetry seen more broadly in the interactions of baryons and mesons. Isospin symmetry remains an important concept in particle physics, and a close examination of this symmetry historically led directly to the discovery and understanding of quarks and of the development of Yang–Mills theory.

3.1   Motivation for isospin

Isospin was introduced by Werner Heisenberg in 1932[3] to explain symmetries of the then newly discovered neutron:

- The mass of the neutron and the proton are almost identical: they are nearly degenerate, and both are thus often called nucleons. Although the proton has a positive charge, and the neutron is neutral, they are almost identical in all other respects.

- The strength of the strong interaction between any pair of nucleons is the same, independent of whether they are interacting as protons or as neutrons.

Thus, isospin was introduced as a concept well before the development in the 1960s of the quark model which provides our modern understanding. The name *isospin* however, was introduced by Eugene Wigner in 1937.[4]

Protons and neutrons, baryons of spin $\frac{1}{2}$, were grouped together as nucleons because they both have nearly the same mass and interact in nearly the same way. Thus, it was convenient to treat them as being different states of the same particle. Since a spin $\frac{1}{2}$ particle has two states, the two were said to be of isospin $\frac{1}{2}$. The proton and neutron were then associated with different isospin projections $I_3 = +\frac{1}{2}$ and $-\frac{1}{2}$ respectively. When constructing a physical theory of nuclear forces, one could then simply assume that it does not depend on isospin.

These considerations would also prove useful in the analysis of meson-nucleon interactions after the discovery of the pions in 1947. The three pions ($\pi+$, $\pi0$, $\pi-$) could be assigned to an isospin triplet with $I = 1$ and $I_3 = +1$, 0 or -1. By assuming that isospin was conserved by nuclear interactions, the new mesons were more easily accommodated by nuclear theory.

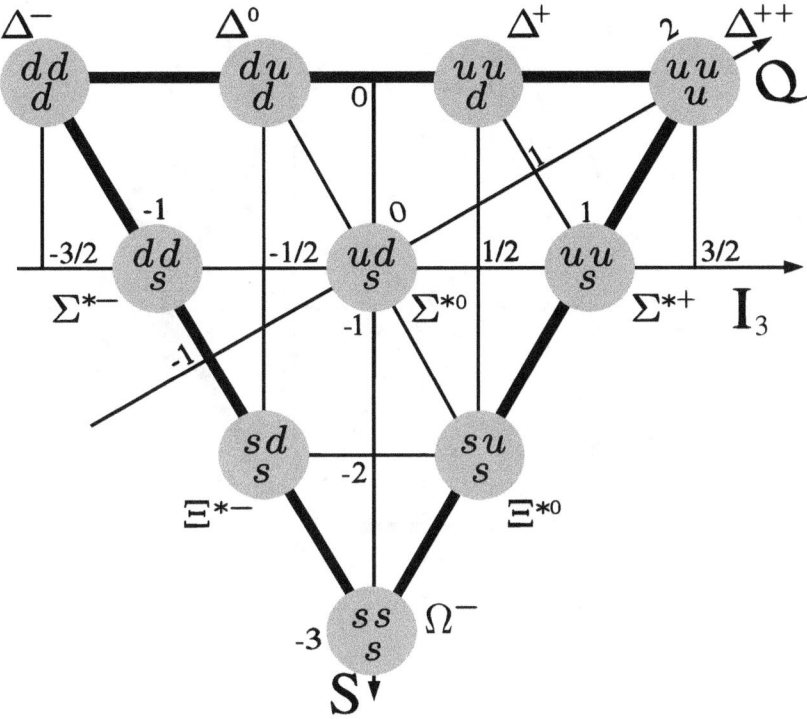

Combinations of three u, d or s-quarks forming baryons with spin-³⁄₂ form the baryon decuplet.

As further particles were discovered, they were assigned into isospin multiplets according to the number of different charge states seen: 2 doublets, $I = -\frac{1}{2}$ and $I = \frac{1}{2}$ of K mesons (K−, K0),(K+, K0), a triplet $I = 1$ of Sigma baryons (Σ+, Σ0, Σ−) a singlet $I = 0$ Lambda baryon (Λ0), a quartet $I = \frac{3}{2}$ Delta baryons (Δ++, Δ+, Δ0, Δ−), and so on. This multiplet structure was combined with strangeness in Murray Gell-Mann's eightfold way, ultimately leading to the quark model and quantum chromodynamics.

3.2 Modern understanding of isospin

Observation of the light baryons (those made of up, down and strange quarks) lead us to believe that some of these particles are so similar in terms of their strong interactions that they can be treated as different states of the same particle. In the modern understanding of quantum chromodynamics, this is because up and down quarks are very similar in mass, and have the same strong interactions. Particles made of the same numbers of up and down quarks have similar masses and are grouped together. For examples, the particles known as the Delta baryons—baryons of spin $\frac{3}{2}$ made of a mix of three up and down quarks—are grouped together because they all have nearly the same mass (approximately 1232 MeV/c^2), and interact in nearly the same way.

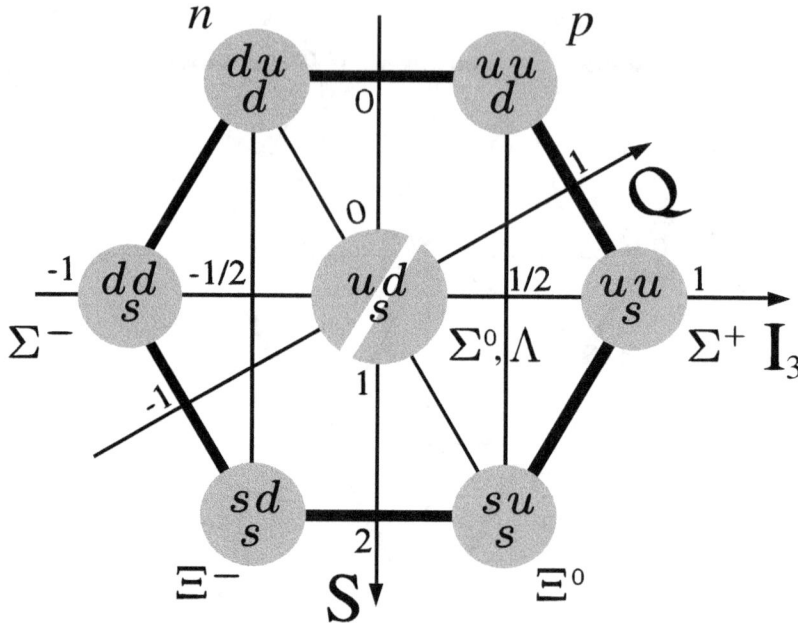

Combinations of three u, d or s-quarks forming baryons with spin-$\frac{1}{2}$ form the baryon octet

However, because the up and down quarks have different charges ($\frac{2}{3}$ e and $-\frac{1}{3}$ e respectively), the four Deltas also have different charges (Δ++ (uuu), Δ+ (uud), Δ0 (udd), Δ− (ddd)). These Deltas could be treated as the same particle and the difference in charge being due to the particle being in different states. Isospin was devised as a parallel to spin to associate an isospin projection (denoted I_3) to each charged state. Since there were four Deltas, four projections were needed. Because isospin was modeled on spin, the isospin projections were made to vary in increments of 1 and to have four increments of 1, you needed an isospin value of $\frac{3}{2}$ (giving the projections $I_3 = \frac{3}{2}$, $\frac{1}{2}$, $-\frac{1}{2}$, $-\frac{3}{2}$). Thus, all the Deltas were said to have isospin $I = \frac{3}{2}$ and each individual charge had different I_3 (e.g. the Δ++ was associated with $I_3 = +\frac{3}{2}$). In the isospin picture, the four Deltas and the two nucleons were thought to be the different states of two particles. In the quark model, the Deltas can be thought of as the excited states of the nucleons.

After the quark model was elaborated, it was noted that the isospin projection was related to the up and down quark content of particles. The relation is

$$I_3 = \frac{1}{2}\left[\left(n_u - n_{\bar{u}}\right) - \left(n_d - n_{\bar{d}}\right)\right]$$

where n_u and n_d are the numbers of up and down quarks respectively, and $n_{\bar{u}}$ and $n_{\bar{d}}$ are the numbers of up and down antiquarks respectively.

By this, the value of I_3 of the nucleons proton (symbol p) and neutron (symbol n) is determined by their quark composition, *uud* for the proton and *udd* for the neutron.

3.3 Isospin symmetry

In quantum mechanics, when a Hamiltonian has a symmetry, that symmetry manifests itself through a set of states that have the same energy; that is, the states are degenerate. In particle physics, the near mass-degeneracy of the neutron and proton points to an approximate symmetry of the Hamiltonian describing the strong interactions. The neutron does have a slightly higher mass due to isospin breaking; this is due to the difference in the masses of the up and down quarks and the effects of the electromagnetic interaction. However, the appearance of an approximate symmetry is still useful, since the small breakings can be described by a perturbation theory, which gives rise to slight differences between the near-degenerate states.

3.3.1 SU(2)

See also: Representation theory of SU(2)

Heisenberg's contribution was to note that the mathematical formulation of this symmetry was in certain respects similar to the mathematical formulation of spin, whence the name "isospin" derives. To be precise, the isospin symmetry is given by the invariance of the Hamiltonian of the strong interactions under the action of the Lie group SU(2). The neutron and the proton are assigned to the doublet (the spin-$\frac{1}{2}$, **2**, or fundamental representation) of SU(2). The pions are assigned to the triplet (the spin-1, **3**, or adjoint representation) of SU(2). Though, there is a difference from the theory of spin: the group action does not preserve flavor.

Like the case for regular spin, the isospin operator $\mathbf{I}$ is vector-valued: it has three components I_x, I_y, I_z which are coordinates in the same 3-dimensional vector space where the **3** representation acts. Note that it has nothing to do with the physical space, except similar mathematical formalism. Isospin is described by two quantum numbers: I, the total isospin, and I_3, an eigenvalue of the I_z projection for which flavor states are eigenstates, not an *arbitrary projection* as in the case of spin. In other words, each I_3 state specifies certain flavor state of a multiplet. The third coordinate (z), to which the "3" subscript refers, is chosen due to notational conventions which relate bases in **2** and **3** representation spaces. Namely, for the spin-$\frac{1}{2}$ case, components of $\mathbf{I}$ are equal to Pauli matrices divided by 2 and $I_z = \frac{1}{2} \tau_3$, where

$$\tau_3 = \begin{pmatrix} 1 & 0 \\ 0 & -1 \end{pmatrix}$$

While the forms of these matrices are the isomorphic to those of spin, *these* Pauli matrices only acts within the Hilbert space of isospin, not that of spin, and therefore is common to denote them with $\boldsymbol{\tau}$ rather than $\boldsymbol{\sigma}$ to avoid confusion.

The power of isospin symmetry and related methods such as the Eightfold Way come from the observation that families of particles with similar masses tend to correspond to the invariant subspaces associated with the irreducible representations of the Lie algebra $\mathfrak{su}(2)$. In this context, an invariant subspace is spanned by basis vectors which correspond to particles in a family. Under the action of the Lie algebra $\mathfrak{su}(2)$, which generates rotations in isospin space, elements corresponding to definite particle states or superpositions of states can be rotated into each other, but can never leave the space (since the subspace is in fact invariant). This is reflective of the symmetry present. The fact that unitary matrices will commute with the Hamiltonian means that the physical quantities calculated do not change even under unitary transformation. In the case of isospin, this machinery is used to reflect the fact that the strong force behaves the same under the exchange of the up and down quark (and by extension the exchange of the proton and the neutron).

3.4 Relationship to flavor

The discovery and subsequent analysis of additional particles, both mesons and baryons, made it clear that the concept of isospin symmetry could be broadened to an even larger symmetry group, now called flavor symmetry. Once the kaons and their property of strangeness became better understood, it started to become clear that these, too, seemed to be a part of an enlarged symmetry that contained isospin as a subgroup. The larger symmetry was named the Eightfold Way by Murray Gell-Mann, and was promptly recognized to correspond to the adjoint representation of SU(3). To better understand the origin of this symmetry, Gell-Mann proposed the existence of up, down and strange quarks which would belong to the fundamental representation of the SU(3) flavor symmetry.

Although isospin symmetry is very slightly broken, SU(3) symmetry is more badly broken, due to the much higher mass of the strange quark compared to the up and down. The discovery of charm, bottomness and topness could lead to further expansions up to SU(6) flavour symmetry, but the very large masses of these quarks makes such symmetries almost useless. In modern applications, such as lattice QCD, isospin symmetry is often treated as exact while the heavier quarks must be treated separately.

3.5   Quark content and isospin

Up and down quarks each have isospin $I = \frac{1}{2}$, and isospin 3-components (I_3) of $\frac{1}{2}$ and $-\frac{1}{2}$ respectively. All other quarks have $I = 0$. In general

$$I_3 = \frac{1}{2}(n_u - n_d).$$

3.5.1   Hadron nomenclature

Main articles: Baryon and Mesons

Hadron nomenclature is based on isospin.[5]

- Particles of isospin $\frac{3}{2}$ can only be made by a mix of three u and d quarks (Delta baryons).

- Particles of isospin 1 are made of a mix of two u and d quarks (Pi mesons, Rho mesons, Sigma baryons with one heavier quark, etc.).

- Particles of isospin $\frac{1}{2}$ can be made of a mix of three u and d quarks (nucleons) or from one u or d quark with heavier quarks (K mesons, D mesons, Xi baryons, etc.)

- Particles of isospin 0 can be made of one u and one d quark (Eta mesons, Omega mesons, Lambda baryons, etc.), or from no u or d quarks at all (Omega baryons, Phi mesons, etc.), with heavier quarks in all cases.

3.5.2   Isospin symmetry of quarks

In the framework of the Standard Model, the isospin symmetry of the proton and neutron are reinterpreted as the isospin symmetry of the up and down quarks. Technically, the nucleon doublet states are seen to be linear combinations of products of 3-particle isospin doublet states and spin doublet states. That is, the (spin-up) proton wave function, in terms of quark-flavour eigenstates, is described by

$$|p\uparrow\rangle = \tfrac{1}{3\sqrt{2}} \left(\ |duu\rangle \quad |udu\rangle \quad |uud\rangle \ \right) \begin{pmatrix} 2 & -1 & -1 \\ -1 & 2 & -1 \\ -1 & -1 & 2 \end{pmatrix} \begin{pmatrix} |\downarrow\uparrow\uparrow\rangle \\ |\uparrow\downarrow\uparrow\rangle \\ |\uparrow\uparrow\downarrow\rangle \end{pmatrix} \text{[6]}$$

and the (spin-up) neutron by

$$|n\uparrow\rangle = \tfrac{1}{3\sqrt{2}} \left(\ |udd\rangle \quad |dud\rangle \quad |ddu\rangle \ \right) \begin{pmatrix} 2 & -1 & -1 \\ -1 & 2 & -1 \\ -1 & -1 & 2 \end{pmatrix} \begin{pmatrix} |\downarrow\uparrow\uparrow\rangle \\ |\uparrow\downarrow\uparrow\rangle \\ |\uparrow\uparrow\downarrow\rangle \end{pmatrix} \text{[6]}$$

Here, $|u\rangle$ is the up quark flavour eigenstate, and $|d\rangle$ is the down quark flavour eigenstate, while $|\uparrow\rangle$ and $|\downarrow\rangle$ are the eigenstates of S_z. Although these superpositions are the technically correct way of denoting a proton and neutron in terms of quark flavour and spin eigenstates, for brevity, they are often simply referred to as "*uud*" and "*udd*". Note also that the derivation above assumes exact isospin symmetry and is modified by SU(2)-breaking terms.

Similarly, the isospin symmetry of the pions are given by:

$$|\pi^+\rangle = |u\overline{d}\rangle$$
$$|\pi^0\rangle = \frac{1}{\sqrt{2}}\left(|u\overline{u}\rangle - |d\overline{d}\rangle\right)$$
$$|\pi^-\rangle = -|d\overline{u}\rangle$$

3.5.3 Weak isospin

Main article: weak isospin

Isospin is similar to, but should not be confused with weak isospin. Briefly, weak isospin is the gauge symmetry of the weak interaction which connects quark and lepton doublets of left-handed particles in all generations; for example, up and down quarks, top and bottom quarks, electrons and electron neutrinos. By contrast (strong) isospin connects only up and down quarks, acts on both chiralities (left and right) and is a global (not a gauge) symmetry.

3.6 Gauged isospin symmetry

Attempts have been made to promote isospin from a global to a local symmetry. In 1954, Chen Ning Yang and Robert Mills suggested that the notion of protons and neutrons, which are continuously rotated into each other by isospin, should be allowed to vary from point to point. To describe this, the proton and neutron direction in isospin space must be defined at every point, giving local basis for isospin. A gauge connection would then describe how to transform isospin along a path between two points.

This Yang–Mills theory describes interacting vector bosons, like the photon of electromagnetism. Unlike the photon, the SU(2) gauge theory would contain self-interacting gauge bosons. The condition of gauge invariance suggests that they have zero mass, just as in electromagnetism.

Ignoring the massless problem, as Yang and Mills did, the theory makes a firm prediction: the vector particle should couple to all particles of a given isospin *universally*. The coupling to the nucleon would be the same as the coupling to the kaons. The coupling to the pions would be the same as the self-coupling of the vector bosons to themselves.

When Yang and Mills proposed the theory, there was no candidate vector boson. J. J. Sakurai in 1960 predicted that there should be a massive vector boson which is coupled to isospin, and predicted that it would show universal couplings. The rho mesons were discovered a short time later, and were quickly identified as Sakurai's vector bosons. The couplings of the rho to the nucleons and to each other were verified to be universal, as best as experiment could measure. The fact that the diagonal isospin current contains part of the electromagnetic current led to the prediction of rho-photon mixing and the concept of vector meson dominance, ideas which led to successful theoretical pictures of GeV-scale photon-nucleus scattering.

Although the discovery of the quarks led to reinterpretation of the rho meson as a vector bound state of a quark and an antiquark, it is sometimes still useful to think of it as the gauge boson of a hidden local symmetry[7]

3.7 References

[1] http://www.thefreedictionary.com/isospin

CHAPTER 3. ISOSPIN

[2] Povh, Bogdan; Klaus, Rith; Scholz, Christoph; Zetsche, Frank (2008) [1993]. "2". *Particles and Nuclei*. p. 21. ISBN 978-3-540-79367-0.

[3] Heisenberg, W. (1932). "Über den Bau der Atomkerne". *Zeitschrift für Physik* (in German) **77**: 1–11. Bibcode:1932ZPhy...77..1H. doi:10.1007/BF01342433.

[4] Wigner, E. (1937). "On the Consequences of the Symmetry of the Nuclear Hamiltonian on the Spectroscopy of Nuclei". *Physical Review* **51** (2): 106–119. Bibcode:1937PhRv...51..106W. doi:10.1103/PhysRev.51.106.

[5] C. Amsler et al.; (Particle Data Group) (2008). "Review of Particle Physics: Naming scheme for hadrons" (PDF). *Physics Letters B* **667**: 1. Bibcode:2008PhLB..667....1P. doi:10.1016/j.physletb.2008.07.018.

[6] Greiner, W.; Müller, B. (1989). *Quantum Mechanics: Symmetries*. Springer-Verlag. p. 279. ISBN 3-540-58080-8.

[7] Bando, M.; Kugo, T.; Uehara, S.; Yamawaki, K.; Yanagida, T. (1985). "Is the ρ Meson a Dynamical Gauge Boson of Hidden Local Symmetry?". *Physical Review Letters* **54** (12): 1215–1218. Bibcode:1985PhRvL..54.1215B. doi:10.1103/PhysRevL15.PMID10030967.

3.8 Further reading

- Itzykson, C.; Zuber, J.-B. (1980). *Quantum Field Theory*. McGraw-Hill. ISBN 0-07-032071-3.

- Griffiths, D. (1987). *Introduction to Elementary Particles*. John Wiley & Sons. ISBN 0-471-60386-4.

3.9 External links

- **Nuclear Structure and Decay Data - IAEA** Nuclides' Isospin

Chapter 4

Charm (quantum number)

Charm (symbol C) is a flavour quantum number representing the difference between the number of charm quarks (c) and charm antiquarks (c) that are present in a particle:

$$C = n_c - n_{\bar{c}}.$$

By convention, the sign of flavour quantum numbers agree with the sign of the electric charge carried by the quark of corresponding flavour. The charm quark, which carries an electric charge (Q) of $+\frac{2}{3}$, therefore carries a charm of $+1$. The charm antiquarks have the opposite charge ($Q = -\frac{2}{3}$), and flavour quantum numbers ($C = -1$).

As with any flavour-related quantum numbers, charm is preserved under strong and electromagnetic interaction, but not under weak interaction (see CKM matrix). For first-order weak decays, that is processes involving only one quark decay, charm can only vary by 1 ($\Delta C = \pm 1, 0$). Since first-order processes are more common than second-order processes (involving two quark decays), this can be used as an approximate "selection rule" for weak decays.

4.1 Further reading

- Lessons in Particle Physics Luis Anchordoqui and Francis Halzen, University of Wisconsin, 18th Dec. 2009

17

Chapter 5

Strangeness

This article is about a concept in particle physics. For the definition of "strangeness", see wikt:strangeness. For other uses, see Strange (disambiguation).

In particle physics, **strangeness** (*"S"*) is a property of particles, expressed as a quantum number, for describing decay of particles in strong and electromagnetic reactions, which occur in a short period of time. The strangeness of a particle is defined as:

$$S = -(n_s - n_{\bar{s}})$$

where n_s represents the number of strange quarks (s) and n_s represents the number of strange antiquarks (s).

The terms *strange* and *strangeness* predate the discovery of the quark, and were adopted after its discovery in order to preserve the continuity of the phrase; strangeness of anti-particles being referred to as +1, and particles as −1 as per the original definition. For all the quark flavor quantum numbers (strangeness, charm, topness and bottomness) the convention is that the flavor charge and the electric charge of a quark have the same sign. With this, any flavor carried by a charged meson has the same sign as its charge.

5.1 Conservation

Strangeness was introduced by Murray Gell-Mann and Kazuhiko Nishijima to explain the fact that certain particles, such as the kaons or certain hyperons, were created easily in particle collisions, yet decayed much more slowly than expected for their large masses and large production cross sections. Noting that collisions seemed to always produce pairs of these particles, it was postulated that a new conserved quantity, dubbed "strangeness", was preserved during their creation, but *not* conserved in their decay.

In our modern understanding, strangeness is conserved during the strong and the electromagnetic interactions, but not during the weak interactions. Consequently, the lightest particles containing a strange quark cannot decay by the strong interaction, and must instead decay via the much slower weak interaction. In most cases these decays change the value of the strangeness by one unit. However, this doesn't necessarily hold in second-order weak reactions, where there are mixes of K0 and K0 mesons. All in all, the amount of strangeness can change in a weak interaction reaction by +1, 0 or −1 (depending on the reaction).

5.2 See also

- Strangeness production

5.3 References

- D.J. Griffiths (1987). *Introduction to Elementary Particles*. John Wiley & Sons. ISBN 0-471-60386-4.

5.4 Further reading

- Lessons in Particle Physics Luis Anchordoqui and Francis Halzen, University of Wisconsin, 18th Dec. 2009

Chapter 6

Topness

Topness (also called **truth**), a flavour quantum number, represents the difference between the number of top quarks (t) and number of top antiquarks (t) that are present in a particle:

$$T = n_t - n_{\bar{t}}$$

By convention, top quarks have a topness of $+1$ and top antiquarks have a topness of -1. The term "topness" is rarely used; most physicists simply refer to "the number of top quarks" and "the number of top antiquarks".

6.1 Conservation

Like all flavour quantum numbers, topness is preserved under strong and electromagnetic interactions, but not under weak interaction. However the top quark is extremely unstable, with a half-life under 10^{-23} s, which is the required time for the strong interaction to take place. For that reason the top quark does not hadronize, that is it never forms any meson or baryon, so the topness of a meson or a baryon is every time equal at zero. By the time it can interact strongly it has already decayed to another flavour of quark (usually to a bottom quark).

6.2 Further reading

- Anchordoqui, L.; Halzen, F. (2009). "Lessons in Particle Physics". arXiv:0906.1271 [physics.ed-ph].

Chapter 7

Bottomness

In physics, **bottomness** (symbol B') also called **beauty**, is a flavour quantum number reflecting the difference between the number of bottom antiquarks ($n_{\bar{b}}$) and the number of bottom quarks (n_b) that are present in a particle:

$$B' = -(n_b - n_{\bar{b}})$$

Bottom quarks have (by convention) a bottomness of -1 while bottom antiquarks have a bottomness of $+1$. The convention is that the flavour quantum number sign for the quark is the same as the sign of the electric charge (symbol Q) of that quark (in this case, $Q = -\frac{1}{3}$).

As with other flavour-related quantum numbers, bottomness is preserved under strong and electromagnetic interactions, but not under weak interactions. For first-order weak reactions, it holds that $\Delta B' = \pm 1$.

This term is rarely used. Most physicists simply refer to "the number of bottom quarks" and "the number of bottom antiquarks".

7.1 Further reading

- Anchordoqui, L.; Halzen, F. (2009). "Lessons in Particle Physics". arXiv:0906.1271 [physics.ed-ph].

Chapter 8

Baryon number

In particle physics, the **baryon number** is a strictly conserved additive quantum number of a system. It is defined as

$$B = \frac{1}{3}\left(n_{\text{q}} - n_{\bar{\text{q}}}\right),$$

where n_{q} is the number of quarks, and $n_{\bar{\text{q}}}$ is the number of antiquarks. Baryons (three quarks) have a baryon number of +1, mesons (one quark, one antiquark) have a baryon number of 0, and antibaryons (three antiquarks) have a baryon number of −1. Exotic hadrons like pentaquarks (four quarks, one antiquark) and tetraquarks (two quarks, two antiquarks) are also classified as baryons and mesons depending on their baryon number.

8.1   Baryon number vs. quark number

See also: Color charge

Quarks carry not only electric charge, but also charges such as color charge and weak isospin. Because of a phenomenon known as *color confinement*, a hadron cannot have a net color charge; that is, the total color charge of a particle has to be zero ("white"). A quark can have one of three "colors", dubbed "red", "green", and "blue".

For normal hadrons, a white color can thus be achieved in one of three ways:

- A quark of one color with an antiquark of the corresponding anticolor, giving a meson with baryon number 0,

- Three quarks of different colors, giving a baryon with baryon number +1,

- Three antiquarks into an antibaryon with baryon number −1.

The baryon number was defined long before the quark model was established, so rather than changing the definitions, particle physicists simply gave quarks one third the baryon number. Nowadays it might be more accurate to speak of the conservation of **quark number**.

In theory, exotic hadrons can be formed by adding pairs of quark and antiquark, provided that each pair has a matching color/anticolor. For example, a pentaquark (four quarks, one antiquark) could have the individual quark colors: red, green, blue, blue, and antiblue.

8.2 Particles not formed of quarks

Particles without any quarks have a baryon number of zero. Such particles include leptons (electron, muon, tau and their neutrinos) and gauge bosons (photon, W and Z bosons, gluons, and the Higgs boson); or the hypothetical graviton.

8.3 Conservation

See also: Conservation law (physics)

The baryon number is conserved in nearly all the interactions of the Standard Model. 'Conserved' means that the sum of the baryon number of all incoming particles is the same as the sum of the baryon numbers of all particles resulting from the reaction. An exception is the chiral anomaly proposed by some extensions of the standard model. However, sphalerons are not all that common. Electroweak sphalerons can only change the baryon number by 3. No experimental evidence of sphalerons has yet been observed.

The still hypothetical idea of a grand unified theory allows for the changing of a baryon into several leptons (see $B - L$), thus violating the conservation of both baryon and lepton numbers.[1] Proton decay would be an example of such a process taking place, but has never been observed.

8.4 See also

- Lepton number
- Flavour (particle physics)
- Isospin
- Hypercharge
- Proton decay
- $B - L$

8.5 References

[1] Griffiths, David (2008). *Introduction to Elementary Particles* (2nd ed.). New York: John Wiley & Sons. p. 77. ISBN 9783527618477. In the grand unified theories new interactions are contemplated, permitting decays such as p+ → e+ + π0 or p+ → ν
μ + π+ in which baryon number and lepton number change.

Chapter 9

Lepton number

In particle physics, the **lepton number** is the number of leptons minus the number of antileptons. In equation form,

$$L = n_\ell - n_{\bar{\ell}}$$

so all leptons have assigned a value of +1, antileptons −1, and non-leptonic particles 0. Lepton number (sometimes also called lepton charge) is an additive quantum number, which means that its sum is preserved in interactions (as opposed to multiplicative quantum numbers such as parity, where the product is preserved instead).

Beside the leptonic number, **leptonic family numbers** are also defined:

- L_e , the **electronic number** for the electron and the electron neutrino;

- $L\mu$, the **muonic number** for the muon and the muon neutrino;

- $L\tau$, the **tauonic number** for the tau and the tau neutrino;

with the same assigning scheme as the leptonic number: +1 for particles of the corresponding family, −1 for the antiparticles, and 0 for leptons of other families or non-leptonic particles.

An example is the muon decay. Like many lepton interactions, muon decay is a Weak Interaction. This is cited as a test for special relativity testing the time dilation effect

9.1 Violations of the lepton number conservation laws

In the Standard Model, leptonic family numbers (LF numbers) would be preserved if neutrinos were massless. Since neutrino oscillations have been observed, neutrinos do have a tiny nonzero mass and conservation laws for LF numbers are therefore only approximate. This means the conservation laws are violated, although because of the smallness of the neutrino mass they still hold to a very large degree for interactions containing charged leptons. However, the (total) lepton number conservation law must still hold (under the Standard Model). Thus, it is possible to see rare muon decays such as μ → eγ or μN→eN:[1]

Because the lepton number conservation law in fact is violated by chiral anomalies, there are problems applying this symmetry universally over all energy scales. However, the quantum number $B - L$ is much more likely to work and is seen in different models such as the Pati–Salam model.

Experiments such as MEGA and SINDRUM have searched for lepton number violation in muon decays to electrons; MEG set the current branching limit of order 10^{-13} and plans to lower to limit to 10^{-14} after 2016. Some BSM theories such as SUSY predict branching ratios of order 10^{-12} to 10^{-14}.[1] The Mu2e experiment in construction has a planned sensitivity of order 10^{-17}.

9.2 References

[1] "New Limit on the Lepton-Flavor-Violating Decay mu to e+gamma". *PRL*. 21 Oct 2011. arXiv:1107.5547. Bibcode:2011PhA. doi:10.1103/PhysRevLett.107.171801.

- Griffiths, David J. (1987). *Introduction to Elementary Particles*. Wiley, John & Sons, Inc. ISBN 0-471-60386-4.

- Tipler, Paul; Llewellyn, Ralph (2002). *Modern Physics (4th ed.)*. W. H. Freeman. ISBN 0-7167-4345-0.

- M. Raidal et al. (2008). *Eur. Phys. J. C 57, 13*. Missing or empty |title= (help)

Chapter 10

Weak isospin

In particle physics, **weak isospin** is a quantum number relating to the weak interaction, and parallels the idea of isospin under the strong interaction. Weak isospin is usually given the symbol T or I with the third component written as T_z, T_3, I_z or I_3.[1] Weak isospin is a complement of the weak hypercharge, which unifies weak interactions with electromagnetic interactions. It can be understood as the eigenvalue of a charge operator.

The **weak isospin conservation law** relates the conservation of T_3; all weak interactions must preserve T_3. It is also conserved by the other interactions and is therefore a conserved quantity in general. For this reason T_3 is more important than T and often the term "weak isospin" refers to the "3rd component of weak isospin".

10.1 Relation with chirality

Fermions with negative chirality (also called left-handed fermions) have $T = \frac{1}{2}$ and can be grouped into doublets with $T_3 = \pm \frac{1}{2}$ that behave the same way under the weak interaction. For example, up-type quarks (u, c, t) have $T_3 = +\frac{1}{2}$ and always transform into down-type quarks (d, s, b), which have $T_3 = -\frac{1}{2}$, and vice versa. On the other hand, a quark never decays weakly into a quark of the same T_3. Something similar happens with left-handed leptons, which exist as doublets containing a charged lepton (e–, μ–, τ–) with $T_3 = -\frac{1}{2}$ and a neutrino (ν
e, ν
μ, ν
τ) with $T_3 = \frac{1}{2}$.

Fermions with positive chirality (also called right-handed fermions) have $T = 0$ and form singlets that do not undergo weak interactions.

Electric charge, Q, is related to weak isospin, T_3, and weak hypercharge, YW, by

$$Q = T_3 + \frac{Y_{\mathrm{W}}}{2}.$$

10.2 Weak isospin and the W bosons

The symmetry associated with spin is SU(2). This requires gauge bosons to transform between weak isospin charges: bosons W+, W– and W0. This implies that W bosons have a $T = 1$, with three different values of T_3.

- W+ boson ($T_3 = +1$) is emitted in transitions $\{(T_3 = +\frac{1}{2}) \rightarrow (T_3 = -\frac{1}{2})\}$,

- W– boson ($T_3 = -1$) is emitted in transitions $\{(T_3 = -\frac{1}{2}) \rightarrow (T_3 = +\frac{1}{2})\}$.

- W0 boson ($T_3 = 0$) would be emitted in reactions where T_3 does not change. However, under electroweak unification, the W0 boson mixes with the weak hypercharge gauge boson B, resulting in the observed Z0 boson and the photon of Quantum Electrodynamics.

10.3 See also

- Field theoretical formulation of standard model
- Weak hypercharge

10.4 References

[1] Ambiguities: I is also used as sign for the 'normal' isospin, same for the third component I_3 aka I_z. T is also used as the sign for Topness. This article uses T and T_3.

Chapter 11

Electric charge

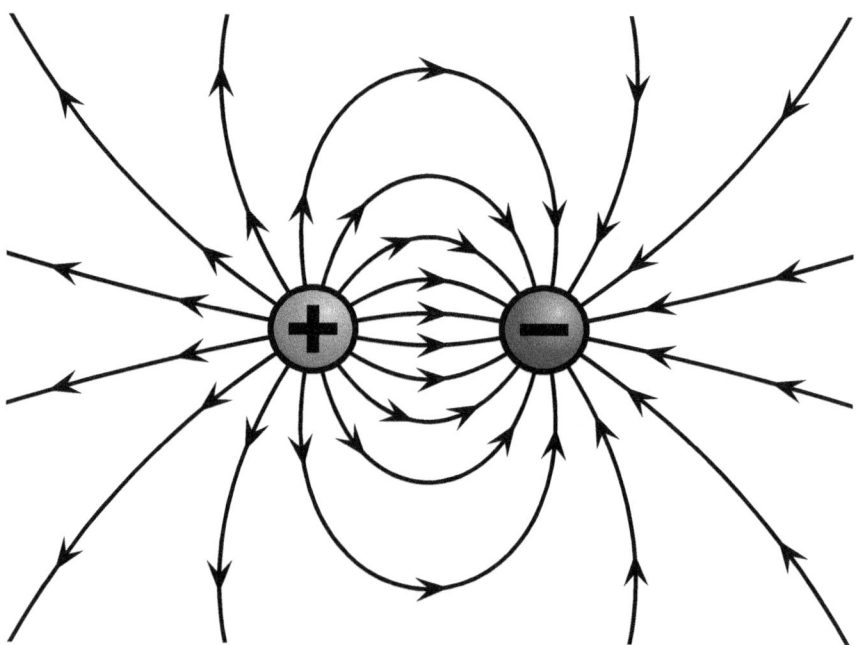

Electric field of a positive and a negative point charge.

Electric charge is the physical property of matter that causes it to experience a force when placed in an electromagnetic field. There are two types of electric charges: positive and negative. Positively charged substances are repelled from other positively charged substances, but attracted to negatively charged substances; negatively charged substances are repelled from negative and attracted to positive. An object is negatively charged if it has an excess of electrons, and is otherwise positively charged or uncharged. The SI derived unit of electric charge is the coulomb (C), although in electrical engineering it is also common to use the ampere-hour (Ah), and in chemistry it is common to use the elementary charge (e) as a unit. The symbol Q is often used to denote charge. The early knowledge of how charged substances interact is now called classical electrodynamics, and is still very accurate if quantum effects do not need to be considered.

The *electric charge* is a fundamental conserved property of some subatomic particles, which determines their electromagnetic interaction. Electrically charged matter is influenced by, and produces, electromagnetic fields. The interaction between a moving charge and an electromagnetic field is the source of the electromagnetic force, which is one of the four fundamental forces (See also: magnetic field).

Twentieth-century experiments demonstrated that electric charge is *quantized*; that is, it comes in integer multiples of individual small units called the elementary charge, e, approximately equal to 1.602×10^{-19} coulombs (except for particles called quarks, which have charges that are integer multiples of $e/3$). The proton has a charge of $+e$, and the electron has a charge of $-e$. The study of charged particles, and how their interactions are mediated by photons, is called quantum electrodynamics.

11.1 Overview

Charge is the fundamental property of forms of matter that exhibit electrostatic attraction or repulsion in the presence of other matter. Electric charge is a characteristic property of many subatomic particles. The charges of free-standing particles are integer multiples of the elementary charge e; we say that electric charge is *quantized*. Michael Faraday, in his electrolysis experiments, was the first to note the discrete nature of electric charge. Robert Millikan's oil-drop experiment demonstrated this fact directly, and measured the elementary charge.

By convention, the charge of an electron is -1, while that of a proton is $+1$. Charged particles whose charges have the same sign repel one another, and particles whose charges have different signs attract. Coulomb's law quantifies the electrostatic force between two particles by asserting that the force is proportional to the product of their charges, and inversely proportional to the square of the distance between them.

The charge of an antiparticle equals that of the corresponding particle, but with opposite sign. Quarks have fractional charges of either $-\frac{1}{3}$ or $+\frac{2}{3}$, but free-standing quarks have never been observed (the theoretical reason for this fact is asymptotic freedom).

The electric charge of a macroscopic object is the sum of the electric charges of the particles that make it up. This charge is often small, because matter is made of atoms, and atoms typically have equal numbers of protons and electrons, in which case their charges cancel out, yielding a net charge of zero, thus making the atom neutral.

An *ion* is an atom (or group of atoms) that has lost one or more electrons, giving it a net positive charge (cation), or that has gained one or more electrons, giving it a net negative charge (anion). *Monatomic ions* are formed from single atoms, while *polyatomic ions* are formed from two or more atoms that have been bonded together, in each case yielding an ion with a positive or negative net charge.

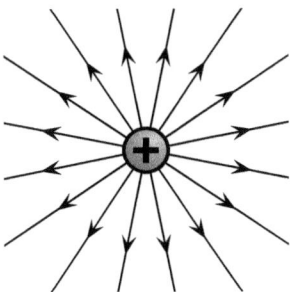

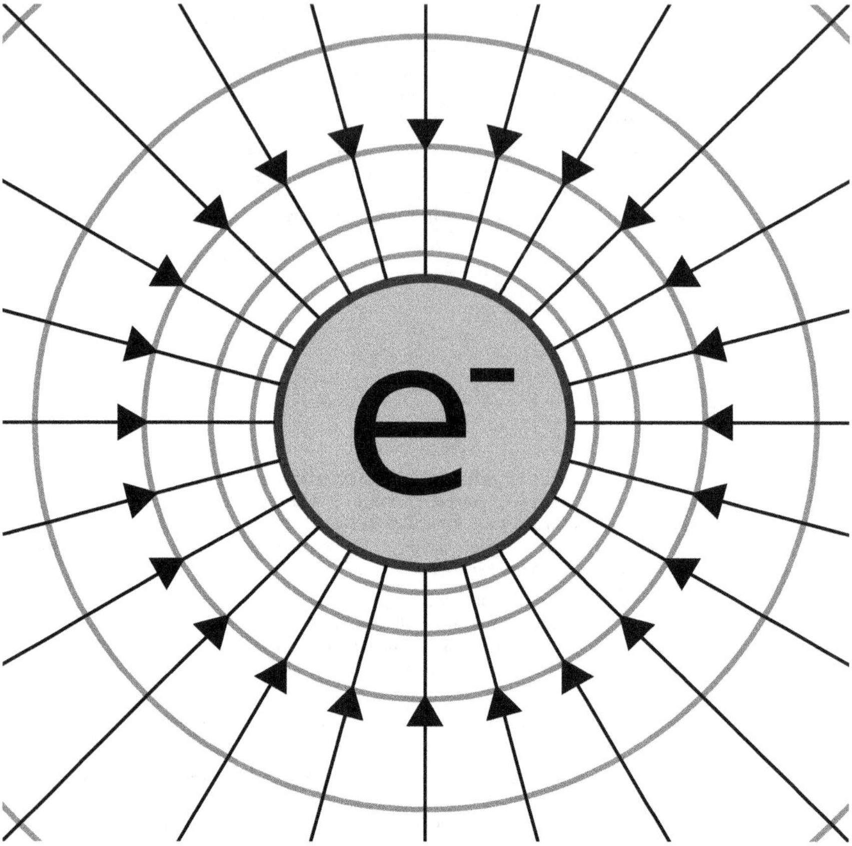

Diagram showing field lines and equipotentials around an electron, a negatively charged particle. In an electrically neutral atom, the number of electrons is equal to the number of protons (which are positively charged), resulting in a net zero overall charge

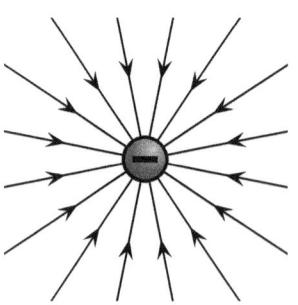

Electric field induced by a positive electric charge (left) and a field induced by a negative electric charge (right).

During formation of macroscopic objects, constituent atoms and ions usually combine to form structures composed of neutral *ionic compounds* electrically bound to neutral atoms. Thus macroscopic objects tend toward being neutral overall, but macroscopic objects are rarely perfectly net neutral.

Sometimes macroscopic objects contain ions distributed throughout the material, rigidly bound in place, giving an overall net positive or negative charge to the object. Also, macroscopic objects made of conductive elements, can more or less easily (depending on the element) take on or give off electrons, and then maintain a net negative or positive charge indefinitely. When the net electric charge of an object is non-zero and motionless, the phenomenon is known as static electricity. This can easily be produced by rubbing two dissimilar materials together, such as rubbing amber with fur or glass with silk. In this way non-conductive materials can be charged to a significant degree, either positively or negatively. Charge taken from one material is moved to the other material, leaving an opposite charge of the same magnitude behind. The law of *conservation of charge* always applies, giving the object from which a negative charge has been taken a positive charge of the same magnitude, and vice versa.

Even when an object's net charge is zero, charge can be distributed non-uniformly in the object (e.g., due to an external electromagnetic field, or bound polar molecules). In such cases the object is said to be polarized. The charge due to polarization is known as bound charge, while charge on an object produced by electrons gained or lost from outside the object is called *free charge*. The motion of electrons in conductive metals in a specific direction is known as electric current.

11.2   Units

The SI unit of quantity of electric charge is the coulomb, which is equivalent to about 6.242×10^{18} e (e is the charge of a proton). Hence, the charge of an electron is approximately -1.602×10^{-19} C. The coulomb is defined as the quantity of charge that has passed through the cross section of an electrical conductor carrying one ampere within one second. The symbol Q is often used to denote a quantity of electricity or charge. The quantity of electric charge can be directly measured with an electrometer, or indirectly measured with a ballistic galvanometer.

After finding the quantized character of charge, in 1891 George Stoney proposed the unit 'electron' for this fundamental unit of electrical charge. This was before the discovery of the particle by J.J. Thomson in 1897. The unit is today treated as nameless, referred to as "elementary charge", "fundamental unit of charge", or simply as "e". A measure of charge should be a multiple of the elementary charge e, even if at large scales charge seems to behave as a real quantity. In some contexts it is meaningful to speak of fractions of a charge; for example in the charging of a capacitor, or in the fractional quantum Hall effect.

In systems of units other than SI such as cgs, electric charge is expressed as combination of only three fundamental quantities such as length, mass and time and not four as in SI where electric charge is a combination of length, mass, time and electric current.

11.3   History

As reported by the ancient Greek mathematician Thales of Miletus around 600 BC, charge (or *electricity*) could be accumulated by rubbing fur on various substances, such as amber. The Greeks noted that the charged amber buttons could attract light objects such as hair. They also noted that if they rubbed the amber for long enough, they could even get an electric spark to jump. This property derives from the triboelectric effect.

In 1600, the English scientist William Gilbert returned to the subject in *De Magnete*, and coined the New Latin word *electricus* from ηλεκτρον (*elektron*), the Greek word for *amber*, which soon gave rise to the English words "electric" and "electricity." He was followed in 1660 by Otto von Guericke, who invented what was probably the first electrostatic generator. Other European pioneers were Robert Boyle, who in 1675 stated that electric attraction and repulsion can act across a vacuum; Stephen Gray, who in 1729 classified materials as conductors and insulators; and C. F. du Fay, who

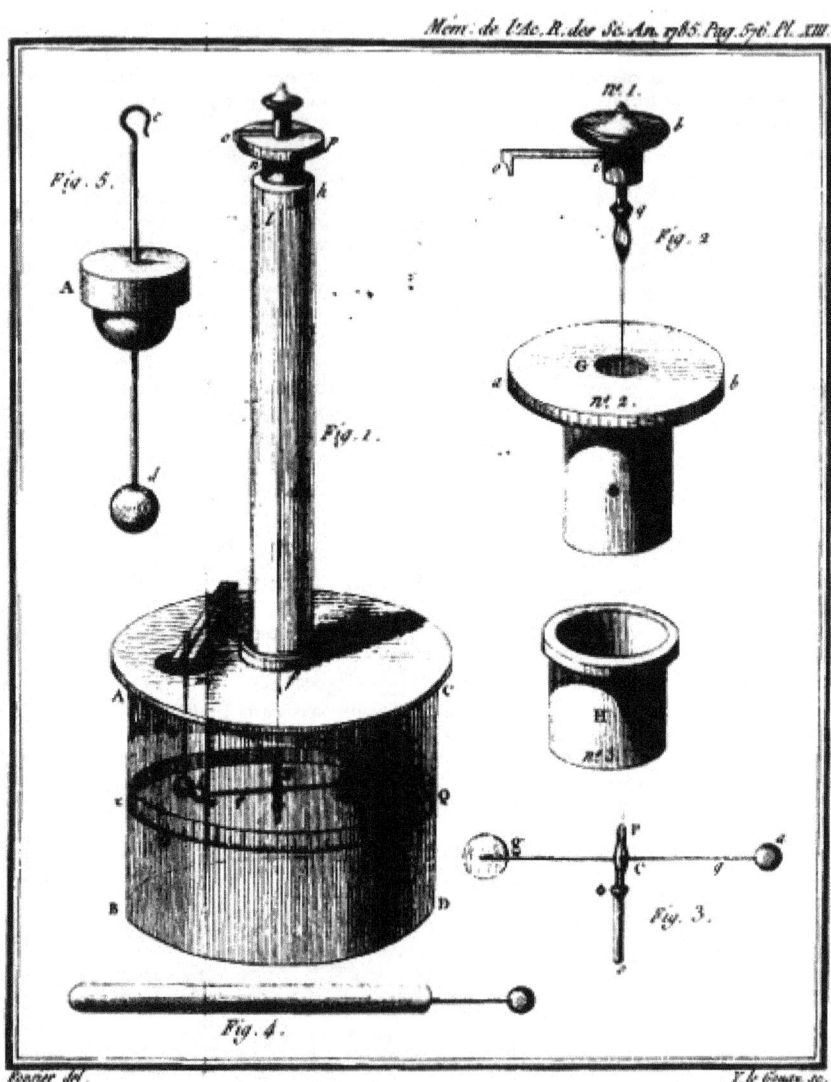

Coulomb's torsion balance

proposed in 1733[1] that electricity comes in two varieties that cancel each other, and expressed this in terms of a two-fluid theory. When glass was rubbed with silk, du Fay said that the glass was charged with *vitreous electricity*, and, when amber was rubbed with fur, the amber was said to be charged with *resinous electricity*. In 1839, Michael Faraday showed that the apparent division between static electricity, current electricity, and bioelectricity was incorrect, and all were a consequence of the behavior of a single kind of electricity appearing in opposite polarities. It is arbitrary which polarity is called positive and which is called negative. Positive charge can be defined as the charge left on a glass rod after being rubbed with silk.[2]

One of the foremost experts on electricity in the 18th century was Benjamin Franklin, who argued in favour of a one-fluid theory of electricity. Franklin imagined electricity as being a type of invisible fluid present in all matter; for example, he believed that it was the glass in a Leyden jar that held the accumulated charge. He posited that rubbing insulating surfaces together caused this fluid to change location, and that a flow of this fluid constitutes an electric current. He also posited that when matter contained too little of the fluid it was "negatively" charged, and when it had an excess it was "positively" charged. For a reason that was not recorded, he identified the term "positive" with vitreous electricity and "negative" with resinous electricity. William Watson arrived at the same explanation at about the same time.

11.4 Static electricity and electric current

Static electricity and electric current are two separate phenomena. They both involve electric charge, and may occur simultaneously in the same object. Static electricity refers to the electric charge of an object and the related electrostatic discharge when two objects are brought together that are not at equilibrium. An electrostatic discharge creates a change in the charge of each of the two objects. In contrast, electric current is the flow of electric charge through an object, which produces no net loss or gain of electric charge.

11.4.1 Electrification by friction

Further information: triboelectric effect

When a piece of glass and a piece of resin—neither of which exhibit any electrical properties—are rubbed together and left with the rubbed surfaces in contact, they still exhibit no electrical properties. When separated, they attract each other.

A second piece of glass rubbed with a second piece of resin, then separated and suspended near the former pieces of glass and resin causes these phenomena:

- The two pieces of glass repel each other.
- Each piece of glass attracts each piece of resin.
- The two pieces of resin repel each other.

This attraction and repulsion is an *electrical phenomena,* and the bodies that exhibit them are said to be *electrified,* or *electrically charged.* Bodies may be electrified in many other ways, as well as by friction. The electrical properties of the two pieces of glass are similar to each other but opposite to those of the two pieces of resin: The glass attracts what the resin repels and repels what the resin attracts.

If a body electrified in any manner whatsoever behaves as the glass does, that is, if it repels the glass and attracts the resin, the body is said to be 'vitreously' electrified, and if it attracts the glass and repels the resin it is said to be 'resinously' electrified. All electrified bodies are found to be either vitreously or resinously electrified.

It is the established convention of the scientific community to define the vitreous electrification as positive, and the resinous electrification as negative. The exactly opposite properties of the two kinds of electrification justify our indicating them by opposite signs, but the application of the positive sign to one rather than to the other kind must be considered as a matter of arbitrary convention, just as it is a matter of convention in mathematical diagram to reckon positive distances towards the right hand.

No force, either of attraction or of repulsion, can be observed between an electrified body and a body not electrified.[3]

Actually, all bodies are electrified, but may appear not to be so by the relative similar charge of neighboring objects in the environment. An object further electrified + or – creates an equivalent or opposite charge by default in neighboring objects, until those charges can equalize. The effects of attraction can be observed in high-voltage experiments, while lower voltage effects are merely weaker and therefore less obvious. The attraction and repulsion forces are codified by Coulomb's Law (attraction falls off at the square of the distance, which has a corollary for acceleration in a gravitational field, suggesting that gravitation may be merely electrostatic phenomenon between relatively weak charges in terms of scale). See also the Casimir effect.

It is now known that the Franklin/Watson model was fundamentally correct. There is only one kind of electrical charge, and only one variable is required to keep track of the amount of charge.[4] On the other hand, just knowing the charge is not a complete description of the situation. Matter is composed of several kinds of electrically charged particles, and these particles have many properties, not just charge.

The most common charge carriers are the positively charged proton and the negatively charged electron. The movement of any of these charged particles constitutes an electric current. In many situations, it suffices to speak of the *conventional current* without regard to whether it is carried by positive charges moving in the direction of the conventional current or by negative charges moving in the opposite direction. This macroscopic viewpoint is an approximation that simplifies electromagnetic concepts and calculations.

At the opposite extreme, if one looks at the microscopic situation, one sees there are many ways of carrying an electric current, including: a flow of electrons; a flow of electron "holes" that act like positive particles; and both negative and positive particles (ions or other charged particles) flowing in opposite directions in an electrolytic solution or a plasma.

Beware that, in the common and important case of metallic wires, the direction of the conventional current is opposite to the drift velocity of the actual charge carriers, i.e., the electrons. This is a source of confusion for beginners.

11.5 Properties

Aside from the properties described in articles about electromagnetism, charge is a relativistic invariant. This means that any particle that has charge Q, no matter how fast it goes, always has charge Q. This property has been experimentally verified by showing that the charge of *one* helium nucleus (two protons and two neutrons bound together in a nucleus and moving around at high speeds) is the same as *two* deuterium nuclei (one proton and one neutron bound together, but moving much more slowly than they would if they were in a helium nucleus).

11.6 Conservation of electric charge

Main article: Charge conservation

The total electric charge of an isolated system remains constant regardless of changes within the system itself. This law is inherent to all processes known to physics and can be derived in a local form from gauge invariance of the wave function. The conservation of charge results in the charge-current continuity equation. More generally, the net change in charge density ϱ within a volume of integration V is equal to the area integral over the current density $\mathbf{J}$ through the closed surface $S = \partial V$, which is in turn equal to the net current I:

$$-\tfrac{d}{dt} \int_V \rho \, dV = \oiint_{\partial V} \mathbf{J} \cdot \mathbf{dS} = \int J dS \cos\theta = I.$$

Thus, the conservation of electric charge, as expressed by the continuity equation, gives the result:

$$I = \frac{dQ}{dt}.$$

The charge transferred between times t_i and t_f is obtained by integrating both sides:

$$Q = \int_{t_i}^{t_f} I \, \mathrm{d}t$$

where I is the net outward current through a closed surface and Q is the electric charge contained within the volume defined by the surface.

11.7 See also

- Quantity of electricity
- SI electromagnetism units

11.8 References

[1] Two Kinds of Electrical Fluid: Vitreous and Resinous – 1733

[2] Electromagnetic Fields (2nd Edition), Roald K. Wangsness, Wiley, 1986. ISBN 0-471-81186-6 (intermediate level textbook)

[3] James Clerk Maxwell *A Treatise on Electricity and Magnetism*, pp. 32-33, Dover Publications Inc., 1954 ASIN: B000HFDK0K, 3rd ed. of 1891

[4] One Kind of Charge

11.9 External links

- How fast does a charge decay?
- Science Aid: Electrostatic charge Easy-to-understand page on electrostatic charge.
- History of the electrical units.

Chapter 12

X (charge)

In particle physics, the **X-charge** (or simply X) is a conserved quantum number associated with the SO(10) grand unification theory.

X is related to the difference between the baryon number B and the lepton number L (that is $B - L$), and the weak hypercharge YW via the relation:

$$X = 5(B - L) - 2Y_W$$

12.1   See also

- Standard Model (mathematical formulation)
- Noether's theorem
- X and Y bosons

12.2   Notes

Chapter 13

Hypercharge

In particle physics, the **hypercharge** (from **hyper**onic + **charge**) Y of a particle is related to the strong interaction, and is distinct from the similarly named weak hypercharge, which has an analogous role in the electroweak interaction. The concept of hypercharge combines and unifies isospin and flavour into a single charge operator.

13.1 Definition

Hypercharge in particle physics is a quantum number relating the strong interactions of the SU(3) model. Isospin is defined in the SU(2) model while the SU(3) model defines hypercharge.

SU(3) weight diagrams (see below) are 2-dimensional with the coordinates referring to two quantum numbers, I_z, which is the z-component of isospin and Y, which is the hypercharge (the sum of strangeness (S), charm (C), bottomness (B'), topness (T), and baryon number (B)). Mathematically, hypercharge is

$$Y = S + C + B' + T + B$$

and conservation of hypercharge implies a conservation of flavour. Strong interactions conserve hypercharge, but weak interactions do not.

13.2 Relation with electric charge and isospin

Main article: Gell-Mann–Nishijima formula

The Gell-Mann–Nishijima formula relates isospin and electric charge

$$Q = I_3 + \frac{1}{2}Y,$$

where I_3 is the third component of isospin and Q is the particle's charge.

Isospin creates multiplets of particles whose average charge is related to the hypercharge by:

$$Y = 2\bar{Q}.$$

since the hypercharge is the same for all members of a multiplet, and the average of the I_3 values is 0.

13.3   SU(3) model in relation to hypercharge

The SU(2) model has multiplets characterized by a quantum number J, which is the total angular momentum. Each multiplet consists of $2J + 1$ substates with equally spaced values of J_z, forming a symmetric arrangement seen in atomic spectra and isospin. This formalises the observation that certain strong baryon decays were not observed, leading to the prediction of the mass, strangeness and charge of the $\Omega-$ baryon.

The SU(3) has *supermultiplets* containing SU(2) multiplets. SU(3) now needs 2 numbers to specify all its sub-states which are denoted by λ_1 and λ_2.

$(\lambda_1 + 1)$ specifies the number of points in the topmost side of the hexagon while $(\lambda_2 + 1)$ specifies the number of points on the bottom side.

13.4   Examples

- The nucleon group (protons with $Q = +1$ and neutrons with $Q = 0$) have an average charge of $+1/2$, so they both have hypercharge $Y = 1$ (baryon number $B = +1$, $S = C = B' = T = 0$). From the Gell-Mann–Nishijima formula we know that proton has isospin $I_3 = +1/2$, while neutron has $I_3 = -1/2$.

- This also works for quarks: for the *up* quark, with a charge of $+2/3$, and an I_3 of $+1/2$, we deduce a hypercharge of $1/3$, due to its baryon number (since you need 3 quarks to make a baryon, a quark has baryon number of $1/3$).

- For a *strange* quark, with charge $-1/3$, a baryon number of $1/3$ and strangeness of -1 we get a hypercharge $Y = -2/3$, so we deduce an $I_3 = 0$. That means that a *strange* quark makes a singlet of its own (same happens with *charm*, *bottom* and *top* quarks), while *up* and *down* constitute an isospin doublet.

13.5   Practical obsolescence

Hypercharge was a concept developed in the 1960s, to organize groups of particles in the *"particle zoo"* and to develop *ad hoc* conservation laws based on their observed transformations. With the advent of the quark model, it is now obvious that (if one only includes the up, down and strange quarks out of the total 6 quarks in the Standard Model), hypercharge Y is the following combination of the numbers of up (n_u), down (n_d), and strange quarks(n_s):

$$Y = \frac{1}{3}(n_u + n_d - 2n_s).$$

In modern descriptions of hadron interaction, it has become more obvious to draw Feynman diagrams that trace through individual quarks composing the interacting baryons and mesons, rather than counting hypercharge quantum numbers. Weak hypercharge, however, remains of practical use in various theories of the electroweak interaction.

13.6   References

- Henry Semat, John R. Albright (1984). *Introduction to atomic and nuclear physics.* Chapman and Hall. ISBN 0-412-15670-9.

Chapter 14

Weak hypercharge

The **weak hypercharge** in particle physics is a quantum number relating the electric charge and the third component of weak isospin. It is conserved (only terms that are overall weak-hypercharge neutral are allowed in the Lagrangian) and is similar to the Gell-Mann–Nishijima formula for the hypercharge of strong interactions (which is not conserved in weak interactions). It is frequently denoted YW and corresponds to the gauge symmetry U(1).[1]

14.1 Definition

Weak hypercharge is the generator of the U(1) component of the electroweak gauge group, SU(2)×U(1) and its associated quantum field B mixes with the W^3 electroweak quantum field to produce the observed Z gauge boson and the photon of quantum electrodynamics.

Weak hypercharge, usually written as YW, satisfies the equality:

$$Q = T_3 + \frac{Y_W}{2}$$

where Q is the electrical charge (in elementary charge units) and T_3 is the third component of weak isospin. Rearranging, the weak hypercharge can be explicitly defined as:

$$Y_W = 2(Q - T_3)$$

Note: sometimes weak hypercharge is scaled so that

$$Y_W = Q - T_3$$

although this is a minority usage.[2]

Hypercharge assignments in the Standard Model are determined up to a twofold ambiguity by demanding cancellation of all anomalies.

14.2 Baryon and lepton number

Weak hypercharge is related to baryon number minus lepton number via:

$X + 2Y_{\mathrm{W}} = 5(B - L)$

where X is a GUT-associated conserved quantum number. Since weak hypercharge is always conserved this implies that baryon number minus lepton number is also always conserved, within the Standard Model and most extensions.

14.2.1   Neutron decay

n → p + e− + ν
e

Hence neutron decay conserves baryon number B and lepton number L separately, so also the difference $B - L$ is conserved.

14.2.2   Proton decay

Proton decay is a prediction of many grand unification theories.

p+ → e+ + π0 → e+ + 2γ

Hence proton decay conserves $B - L$, even though it violates both lepton number and baryon number conservation.

14.3   See also

- Standard Model (mathematical formulation)

14.4   Notes

[1] J. F. Donoghue, E. Golowich, B. R. Holstein (1994). *Dynamics of the standard model.* Cambridge University Press. p. 52. ISBN 0-521-47652-6.

[2] M. R. Anderson (2003). *The mathematical theory of cosmic strings.* CRC Press. p. 12. ISBN 0-7503-0160-0.

Chapter 15

Cabibbo–Kobayashi–Maskawa matrix

In the Standard Model of particle physics, the **Cabibbo–Kobayashi–Maskawa matrix** (**CKM matrix, quark mixing matrix**, sometimes also called **KM matrix**) is a unitary matrix which contains information on the strength of flavour-changing weak decays. Technically, it specifies the mismatch of quantum states of quarks when they propagate freely and when they take part in the weak interactions. It is important in the understanding of CP violation. This matrix was introduced for three generations of quarks by Makoto Kobayashi and Toshihide Maskawa, adding one generation to the matrix previously introduced by Nicola Cabibbo. This matrix is also an extension of the GIM mechanism, which only includes two of the three current families of quarks.

15.1   The matrix

In 1963, Nicola Cabibbo introduced the Cabibbo angle (θ_c) to preserve the universality of the weak interaction.[1] Cabibbo was inspired by previous work by Murray Gell-Mann and Maurice Lévy,[2] on the effectively rotated nonstrange and strange vector and axial weak currents, which he references.[3]

In light of current knowledge (quarks were not yet theorized), the Cabibbo angle is related to the relative probability that down and strange quarks decay into up quarks ($|V_{ud}|^2$ and $|V_{us}|^2$ respectively). In particle physics parlance, the object that couples to the up quark via charged-current weak interaction is a superposition of down-type quarks, here denoted by d'.[4] Mathematically this is:

$$d' = V_{ud}d + V_{us}s,$$

or using the Cabibbo angle:

$$d' = \cos\theta_c d + \sin\theta_c s.$$

Using the currently accepted values for $|V_{ud}|$ and $|V_{us}|$ (see below), the Cabibbo angle can be calculated using

$$\tan\theta_c = \frac{|V_{us}|}{|V_{ud}|} = \frac{0.22534}{0.97427} \rightarrow \theta_c = 13.02°.$$

When the charm quark was discovered in 1974, it was noticed that the down and strange quark could decay into either the up or charm quark, leading to two sets of equations:

$$d' = V_{ud}d + V_{us}s;$$

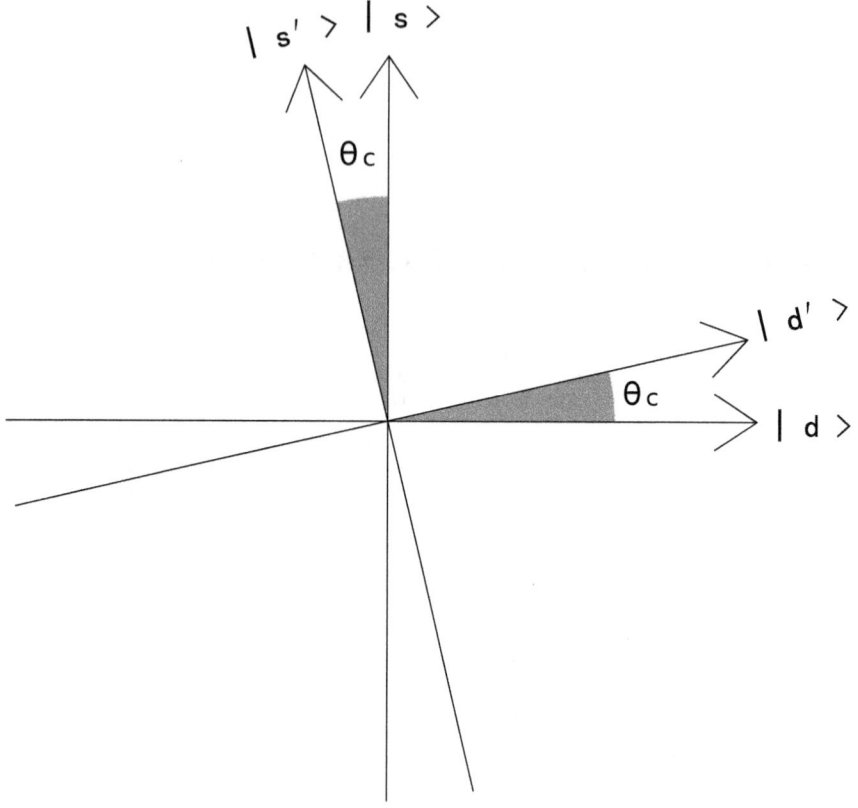

The Cabibbo angle represents the rotation of the mass eigenstate vector space formed by the mass eigenstates $|d\rangle$, $|s\rangle$ into the weak eigenstate vector space formed by the weak eigenstates $|d'\rangle$, $|s'\rangle$. $\theta C = 13.02°$.

$s' = V_{cd}d + V_{cs}s,$

or using the Cabibbo angle:

$d' = \cos\theta_c d + \sin\theta_c s;$

$s' = -\sin\theta_c d + \cos\theta_c s.$

This can also be written in matrix notation as:

$$\begin{bmatrix} d' \\ s' \end{bmatrix} = \begin{bmatrix} V_{ud} & V_{us} \\ V_{cd} & V_{cs} \end{bmatrix} \begin{bmatrix} d \\ s \end{bmatrix},$$

or using the Cabibbo angle

$$\begin{bmatrix} d' \\ s' \end{bmatrix} = \begin{bmatrix} \cos\theta_c & \sin\theta_c \\ -\sin\theta_c & \cos\theta_c \end{bmatrix} \begin{bmatrix} d \\ s \end{bmatrix},$$

where the various $|V_{ij}|^2$ represent the probability that the quark of j flavor decays into a quark of i flavor. This 2×2 rotation matrix is called the Cabibbo matrix.

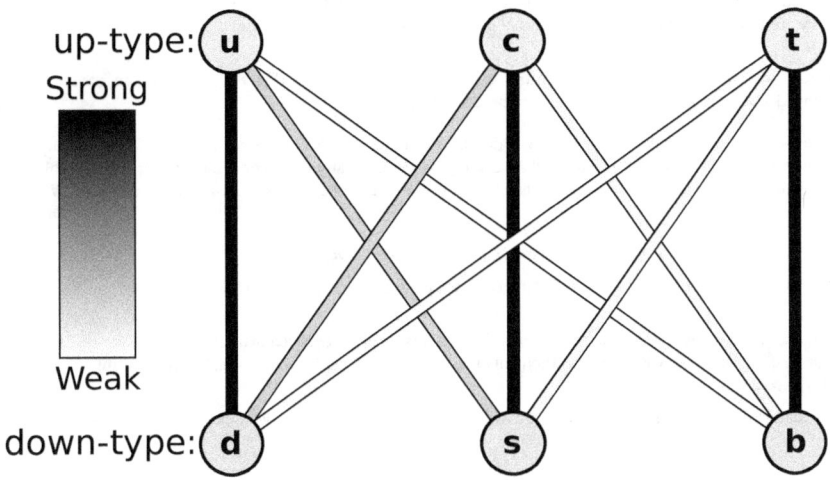

A pictorial representation of the six quarks' decay modes, with mass increasing from left to right.

Observing that CP-violation could not be explained in a four-quark model, Kobayashi and Maskawa generalized the Cabbibo matrix into the Cabibbo–Kobayashi–Maskawa matrix (or CKM matrix) to keep track of the weak decays of three generations of quarks:[5]

$$\begin{bmatrix} d' \\ s' \\ b' \end{bmatrix} = \begin{bmatrix} V_{ud} & V_{us} & V_{ub} \\ V_{cd} & V_{cs} & V_{cb} \\ V_{td} & V_{ts} & V_{tb} \end{bmatrix} \begin{bmatrix} d \\ s \\ b \end{bmatrix}.$$

On the left is the weak interaction doublet partners of up-type quarks, and on the right is the CKM matrix along with a vector of mass eigenstates of down-type quarks. The CKM matrix describes the probability of a transition from one quark i to another quark j. These transitions are proportional to $|V_{ij}|^2$.

Currently, the best determination of the magnitudes of the CKM matrix elements is:[6]

$$\begin{bmatrix} |V_{ud}| & |V_{us}| & |V_{ub}| \\ |V_{cd}| & |V_{cs}| & |V_{cb}| \\ |V_{td}| & |V_{ts}| & |V_{tb}| \end{bmatrix} = \begin{bmatrix} 0.97427 \pm 0.00015 & 0.22534 \pm 0.00065 & 0.00351^{+0.00015}_{-0.00014} \\ 0.22520 \pm 0.00065 & 0.97344 \pm 0.00016 & 0.0412^{+0.0011}_{-0.0005} \\ 0.00867^{+0.00029}_{-0.00031} & 0.0404^{+0.0011}_{-0.0005} & 0.999146^{+0.000021}_{-0.000046} \end{bmatrix}.$$

Note that the choice of usage of down-type quarks in the definition is purely arbitrary and does not represent some sort of deep physical asymmetry between up-type and down-type quarks. We could just as easily define the matrix the other way around, describing weak interaction partners of mass eigenstates of up-type quarks, u', c' and t', in terms of u, c, and t. Since the CKM matrix is unitary (and therefore its inverse is the same as its conjugate transpose), we would obtain essentially the same matrix.

15.2   Counting

To proceed further, it is necessary to count the number of parameters in this matrix, V which appear in experiments, and therefore are physically important. If there are N generations of quarks ($2N$ flavours) then

- An $N \times N$ unitary matrix (that is, a matrix V such that $VV^\dagger = I$, where $V^\dagger$ is the conjugate transpose of V and I is the identity matrix) requires N^2 real parameters to be specified.

- $2N - 1$ of these parameters are not physically significant, because one phase can be absorbed into each quark field (both of the mass eigenstates, and of the weak eigenstates), but an overall common phase is unobservable. Hence, the total number of free variables independent of the choice of the phases of basis vectors is $N^2 - (2N - 1) = (N - 1)^2$.

  - Of these, $N(N - 1)/2$ are rotation angles called quark *mixing angles*.
  - The remaining $(N - 1)(N - 2)/2$ are complex phases, which cause CP violation.

For the case $N = 2$, there is only one parameter which is a mixing angle between two generations of quarks. Historically, this was the first version of CKM matrix when only two generations were known. It is called the **Cabibbo angle** after its inventor Nicola Cabibbo.

For the Standard Model case ($N = 3$), there are three mixing angles and one CP-violating complex phase.[7]

15.3   Observations and predictions

Cabibbo's idea originated from a need to explain two observed phenomena:

1. the transitions $u \leftrightarrow d$, $e \leftrightarrow \nu e$, and $\mu \leftrightarrow \nu\mu$ had similar amplitudes.

2. the transitions with change in strangeness $\Delta S = 1$ had amplitudes equal to 1/4 of those with $\Delta S = 0$.

Cabibbo's solution consisted of postulating weak universality to resolve the first issue, along with a mixing angle θ_c, now called the *Cabibbo angle*, between the d and s quarks to resolve the second.

For two generations of quarks, there are no CP violating phases, as shown by the counting of the previous section. Since CP violations were seen in neutral kaon decays already in 1964, the emergence of the Standard Model soon after was a clear signal of the existence of a third generation of quarks, as pointed out in 1973 by Kobayashi and Maskawa. The discovery of the bottom quark at Fermilab (by Leon Lederman's group) in 1976 therefore immediately started off the search for the missing third-generation quark, the top quark.

Note, however, that the specific values of the angles are *not* a prediction of the standard model: they are open, unfixed parameters. At this time, there is no generally accepted theory that explains why the measured values are what they are.

15.4   Weak universality

The constraints of unitarity of the CKM-matrix on the diagonal terms can be written as

$$\sum_k |V_{ik}|^2 = \sum_i |V_{ik}|^2 = 1$$

for all generations i. This implies that the sum of all couplings of any of the up-type quarks to all the down-type quarks is the same for all generations. This relation is called *weak universality* and was first pointed out by Nicola Cabibbo in 1967. Theoretically it is a consequence of the fact that all SU(2) doublets couple with the same strength to the vector bosons of weak interactions. It has been subjected to continuing experimental tests.

15.5 The unitarity triangles

The remaining constraints of unitarity of the CKM-matrix can be written in the form

$$\sum_k V_{ik} V_{jk}^* = 0.$$

For any fixed and different i and j, this is a constraint on three complex numbers, one for each k, which says that these numbers form the sides of a triangle in the complex plane. There are six choices of i and j (three independent), and hence six such triangles, each of which is called a *unitary triangle*. Their shapes can be very different, but they all have the same area, which can be related to the CP violating phase. The area vanishes for the specific parameters in the Standard Model for which there would be no CP violation. The orientation of the triangles depend on the phases of the quark fields.

Since the three sides of the triangles are open to direct experiment, as are the three angles, a class of tests of the Standard Model is to check that the triangle closes. This is the purpose of a modern series of experiments under way at the Japanese BELLE and the American BaBar experiments, as well as at LHCb in CERN, Switzerland.

15.6 Parameterizations

Four independent parameters are required to fully define the CKM matrix. Many parameterizations have been proposed, and three of the most common ones are shown below.

15.6.1 KM parameters

The original parameterization of Kobayashi and Maskawa used three angles (θ_1, θ_2, θ_3) and a CP-violating phase (δ).[5] Cosines and sines of the angles are denoted ci and si, respectively. θ_1 is the Cabibbo angle.

$$\begin{bmatrix} c_1 & -s_1 c_3 & -s_1 s_3 \\ s_1 c_2 & c_1 c_2 c_3 - s_2 s_3 e^{i\delta} & c_1 c_2 s_3 + s_2 c_3 e^{i\delta} \\ s_1 s_2 & c_1 s_2 c_3 + c_2 s_3 e^{i\delta} & c_1 s_2 s_3 - c_2 c_3 e^{i\delta} \end{bmatrix}.$$

15.6.2 "Standard" parameters

A "standard" parameterization of the CKM matrix uses three Euler angles (θ_{12}, θ_{23}, θ_{13}) and one CP-violating phase (δ_{13}).[8] Couplings between quark generation i and j vanish if $\theta ij = 0$. Cosines and sines of the angles are denoted cij and sij, respectively. θ_{12} is the Cabibbo angle.

$$\begin{bmatrix} 1 & 0 & 0 \\ 0 & c_{23} & s_{23} \\ 0 & -s_{23} & c_{23} \end{bmatrix} \begin{bmatrix} c_{13} & 0 & s_{13}e^{-i\delta_{13}} \\ 0 & 1 & 0 \\ -s_{13}e^{i\delta_{13}} & 0 & c_{13} \end{bmatrix} \begin{bmatrix} c_{12} & s_{12} & 0 \\ -s_{12} & c_{12} & 0 \\ 0 & 0 & 1 \end{bmatrix}$$

$$= \begin{bmatrix} c_{12}c_{13} & s_{12}c_{13} & s_{13}e^{-i\delta_{13}} \\ -s_{12}c_{23} - c_{12}s_{23}s_{13}e^{i\delta_{13}} & c_{12}c_{23} - s_{12}s_{23}s_{13}e^{i\delta_{13}} & s_{23}c_{13} \\ s_{12}s_{23} - c_{12}c_{23}s_{13}e^{i\delta_{13}} & -c_{12}s_{23} - s_{12}c_{23}s_{13}e^{i\delta_{13}} & c_{23}c_{13} \end{bmatrix}.$$

The currently best known values for the standard parameters are:[9]

θ_{12} = 13.04±0.05°, θ_{13} = 0.201±0.011°, θ_{23} = 2.38±0.06°, and δ_{13} = 1.20±0.08 rad.

15.6.3   Wolfenstein parameters

A third parameterization of the CKM matrix was introduced by Lincoln Wolfenstein with the four parameters λ, A, ϱ, and η.[10] The four Wolfenstein parameters have the property that all are of order 1 and are related to the "standard" parameterization:

$\lambda = s_{12}$

$A\lambda^2 = s_{23}$

$A\lambda^3(\varrho - i\eta) = s_{13}e^{-i\delta}$

The Wolfenstein parameterization of the CKM matrix, is an approximation of the standard parameterization. To order λ^3, it is:

$$\begin{bmatrix} 1 - \lambda^2/2 & \lambda & A\lambda^3(\rho - i\eta) \\ -\lambda & 1 - \lambda^2/2 & A\lambda^2 \\ A\lambda^3(1 - \rho - i\eta) & -A\lambda^2 & 1 \end{bmatrix}.$$

The CP violation can be determined by measuring $\varrho - i\eta$.

Using the values of the previous section for the CKM matrix, the best determination of the Wolfenstein parameters is:[11]

λ = 0.2257+0.0009
−0.0010, A = 0.814+0.021
−0.022, ϱ = 0.135+0.031
−0.016, and η = 0.349+0.015
−0.017.

15.7   Nobel Prize

In 2008, Kobayashi and Maskawa shared one half of the Nobel Prize in Physics "for the discovery of the origin of the broken symmetry which predicts the existence of at least three families of quarks in nature".[12] Some physicists were reported to harbor bitter feelings about the fact that the Nobel Prize committee failed to reward the work of Cabibbo, whose prior work was closely related to that of Kobayashi and Maskawa.[13] Asked for a reaction on the prize, Cabibbo preferred to give no comment.[14]

15.8   See also

- Formulation of the Standard Model and CP violations.

- Quantum chromodynamics, flavour and strong CP problem.

- Weinberg angle, a similar angle for Z and photon mixing.

- Pontecorvo–Maki–Nakagawa–Sakata matrix, the equivalent mixing matrix for neutrinos.

- Koide formula

15.9   References

[1] N.Cabibbo(1963). "Unitary Symmetry and Leptonic Decays".*Physical Review Letters***10**(12):531–533.Bibcode:1963PhR1C. doi:10.1103/PhysRevLett.10.531.

[2] M.Gell-Mann,M.Lévy(1960). "The Axial Vector Current in Beta Decay".*Il Nuovo Cimento***16**(4):705–726.doi:10.19738.

[3] L. Maiani (2009). "Sul Premio Nobel Per La Fisica 2008" (PDF). *Il Nuovo Saggiatore* **25** (1–2): 78.

[4] I.S. Hughes (1991). "Chapter 11.1 – Cabibbo Mixing". *Elementary Particles* (3rd ed.). Cambridge University Press. pp. 242–243. ISBN 0-521-40402-9.

[5] M. Kobayashi, T. Maskawa; Maskawa (1973). "CP-Violation in the Renormalizable Theory of Weak Interaction". *Progress of Theoretical Physics* **49** (2): 652–657. Bibcode:1973PThPh..49..652K. doi:10.1143/PTP.49.652.

[6] J. Beringer; Arguin, J. -F.; Barnett, R. M.; Copic, K.; Dahl, O.; Groom, D. E.; Lin, C. -J.; Lys, J.; Murayama, H.; Wohl, C. G.; Yao, W. -M.; Zyla, P. A.; Amsler, C.; Antonelli, M.; Asner, D. M.; Baer, H.; Band, H. R.; Basaglia, T.; Bauer, C. W.; Beatty, J. J.; Belousov, V. I.; Bergren, E.; Bernardi, G.; Bertl, W.; Bethke, S.; Bichsel, H.; Biebel, O.; Blucher, E.; Blusk, S. et al. (2012). "Review of Particles Physics: The CKM Quark-Mixing Matrix" (PDF). *Physical Review D* **80** (1): 1–1526 [162]. Bibcode:2012PhRvD..86a0001B. doi:10.1103/PhysRevD.86.010001.

[7] J.C. Baez (6 March 2005). "Neutrinos and the Mysterious Maki-Nakagawa-Sakata Matrix". Retrieved 2009-01-04. In fact, the Maki–Nakagawa–Sakata matrix actually affects the behavior of all leptons, not just neutrinos. Furthermore, a similar trick works for quarks – but then the matrix U is called the Cabibbo–Kobayashi–Maskawa matrix.

[8] L.L. Chau and W.-Y. Keung (1984). "Comments on the Parametrization of the Kobayashi-Maskawa Matrix". *Physical Review Letters* **53** (19): 1802. Bibcode:1984PhRvL..53.1802C. doi:10.1103/PhysRevLett.53.1802.

[9] Values obtained from values of Wolfenstein parameters in the 2008 *Review of Particle Physics*.

[10] L.Wolfenstein(1983). "Parametrization of the Kobayashi-Maskawa Matrix".*Physical Review Letters***51**(21):19451.1945W. doi:10.1103/PhysRevLett.51.1945.

[11] C. Amsler; Doser, M.; Antonelli, M.; Asner, D.M.; Babu, K.S.; Baer, H.; Band, H.R.; Barnett, R.M.; Bergren, E.; Beringer, J.; Bernardi, G.; Bertl, W.; Bichsel, H.; Biebel, O.; Bloch, P.; Blucher, E.; Blusk, S.; Cahn, R.N.; Carena, M.; Caso, C.; Ceccucci, A.; Chakraborty, D.; Chen, M.-C.; Chivukula, R.S.; Cowan, G.; Dahl, O.; d'Ambrosio, G.; Damour, T.; De Gouvêa, A. et al. (2008). "Review of Particles Physics: The CKM Quark-Mixing Matrix" (PDF). *Physics Letters B* **667**: 1–1340. Bibcode:2008PhLB..667....1P. doi:10.1016/j.physletb.2008.07.018.

[12] "The Nobel Prize in Physics 2008" (Press release). The Nobel Foundation. 7 October 2008. Retrieved 2009-11-24.

[13] V. Jamieson (7 October 2008). "Physics Nobel Snubs key Researcher". *New Scientist*. Retrieved 2009-11-24.

[14] "Nobel, l'amarezza dei fisici italiani". *Corriere della Sera* (in Italian). 7 October 2008. Retrieved 2009-11-24.

15.10   Further reading

- D.J Griffiths (2008). *Introduction to Elementary Particles* (2nd ed.). John Wiley & Sons. ISBN 978-3-527-40601-2.

- B. Povh et al. (1995). *Particles and Nuclei: An Introduction to the Physical Concepts.* Springer. ISBN 3-540-20168-8.

- I.I. Bigi, A.I. Sanda (2000). *CP violation.* Cambridge University Press. ISBN 0-521-44349-0.

- Particle Data Group: The CKM quark-mixing matrix

- Particle Data Group: CP violation in meson decays

- The Babar experiment at SLAC and the BELLE experiment at KEK Japan

Chapter 16

Pontecorvo–Maki–Nakagawa–Sakata matrix

In particle physics, the **Pontecorvo–Maki–Nakagawa–Sakata matrix** (**PMNS matrix**), **Maki–Nakagawa–Sakata matrix** (**MNS matrix**), **lepton mixing matrix**, or **neutrino mixing matrix**, is a unitary matrix[note 1] which contains information on the mismatch of quantum states of neutrinos when they propagate freely and when they take part in the weak interactions. It is important in the understanding of neutrino oscillation. This matrix was introduced in 1962 by Ziro Maki, Masami Nakagawa and Shoichi Sakata,[1] to explain the neutrino oscillations predicted by Bruno Pontecorvo.[2]

16.1   The PMNS matrix

The Standard Model of particle physics contains three generations or "flavors" of neutrinos, ν_e, $\nu\mu$, and $\nu\tau$ labeled according to the charged leptons with which they partner in the charged-current weak interaction. These three eigenstates of the weak interaction form a complete, orthonormal basis for the Standard Model neutrino. Similarly, one can construct an eigenbasis out of three neutrino states of definite mass, ν_1, ν_2, and ν_3, which diagonalize the neutrino's free-particle Hamiltonian. Observations of neutrino oscillation have experimentally determined that for neutrinos, like the quarks, these two eigenbases are not the same - they are "rotated" relative to each other. Each flavor state can thus be written as a superposition of mass eigenstates, and vice-versa. The PMNS matrix, with components Uai corresponding to the amplitude of mass eigenstate i in flavor a, parameterizes the unitary transformation between the two bases:

$$\begin{bmatrix} \nu_e \\ \nu_\mu \\ \nu_\tau \end{bmatrix} = \begin{bmatrix} U_{e1} & U_{e2} & U_{e3} \\ U_{\mu 1} & U_{\mu 2} & U_{\mu 3} \\ U_{\tau 1} & U_{\tau 2} & U_{\tau 3} \end{bmatrix} \begin{bmatrix} \nu_1 \\ \nu_2 \\ \nu_3 \end{bmatrix}.$$

The vector on the left represents a generic neutrino state expressed in the flavor basis, and on the right is the PMNS matrix multiplied by a vector representing the same neutrino state in the mass basis. A neutrino of a given flavor α is thus a "mixed" state of neutrinos with different mass: if one could measure directly that neutrino's mass, it would be found to have mass m_i with probability $|U\alpha i|^2$.

The PMNS matrix for antineutrinos is identical to the matrix for neutrinos under CPT symmetry.

Due to the difficulties of detecting neutrinos, it is much more difficult to determine the individual coefficients than in the equivalent matrix for the quarks (the CKM matrix).

16.1.1   Assumptions

As noted above, PMNS matrix is unitary (i.e. the sum of the square of the values in each row and in each column, which represent the probabilities of different possible events given the same starting point, add up to 100%) in the simplest Standard Model case in which there are three generations of neutrinos with Dirac mass that oscillate between three neutrino mass eigenvalues, an assumption that is made when best fit values for its parameters are calculated.

The PMNS matrix is not necessarily unitary and additional parameters are necessary to describe all possible neutrino mixing parameters, in other models of neutrino oscillation and mass generation, such as the see-saw model, and in general, in the case of neutrinos that have Majorana mass rather than Dirac mass.

There are also additional mass parameters and mixing angles in a simple extension of the PMNS matrix in which there are more than three flavors of neutrinos, regardless of the character of neutrino mass. As of July 2014, scientists studying neutrino oscillation are actively considering fits of the experimental neutrino oscillation data to an extended PMNS matrix with a fourth, light "sterile" neutrino and four mass eigenvalues, although the current experimental data tends to disfavor that possibility.[3][4][5]

16.1.2   Parameterization

In general, there are nine degrees of freedom in any three by three matrix, and in the PMNS matrix, because it is a matrix whose directly physically observable values (the square of the respective entries) are real numbers between zero and 1 form a unitary matrix, the matrix can thus be fully described by four free parameters from which all physically observable properties of the matrix can be discerned.[6] The PMNS matrix is most commonly parameterized by three mixing angles (θ_{12}, θ_{23} and θ_{13}) and a single phase called δCP related to charge-parity violations (i.e. differences in the rates of oscillation between two states with opposite starting points which makes the order in time in which events take place necessary to predict their oscillation rates), in which case the matrix can be written as:

$$
\begin{bmatrix} 1 & 0 & 0 \\ 0 & c_{23} & s_{23} \\ 0 & -s_{23} & c_{23} \end{bmatrix} \begin{bmatrix} c_{13} & 0 & s_{13}e^{-i\delta_{CP}} \\ 0 & 1 & 0 \\ -s_{13}e^{i\delta_{CP}} & 0 & c_{13} \end{bmatrix} \begin{bmatrix} c_{12} & s_{12} & 0 \\ -s_{12} & c_{12} & 0 \\ 0 & 0 & 1 \end{bmatrix}
$$
$$
= \begin{bmatrix} c_{12}c_{13} & s_{12}c_{13} & s_{13}e^{-i\delta_{CP}} \\ -s_{12}c_{23} - c_{12}s_{23}s_{13}e^{i\delta_{CP}} & c_{12}c_{23} - s_{12}s_{23}s_{13}e^{i\delta_{CP}} & s_{23}c_{13} \\ s_{12}s_{23} - c_{12}c_{23}s_{13}e^{i\delta_{CP}} & -c_{12}s_{23} - s_{12}c_{23}s_{13}e^{i\delta_{CP}} & c_{23}c_{13} \end{bmatrix}.
$$

where s_{ij} and c_{ij} are used to denote $\sin\theta_{ij}$ and $\cos\theta_{ij}$ respectively. In the case of Majorana neutrinos, two extra complex phases are needed, as the phase of Majorana fields cannot be freely redefined due to the condition $\nu = \nu^c$. An infinite number of possible parameterizations exist; one other common example being the Wolfenstein parameterization.

The mixing angles have been measured by a variety of experiments (see neutrino mixing for a description). The CP-violating phase δCP has not been measured directly, but estimates can be obtained by fits using the other measurements.

16.1.3   Experimentally measured parameter values

As of July 2014, the current best directly measured values are:[7][8]

$$\sin^2 2\theta_{12} = 0.857 \pm 0.024$$
$$\sin^2 2\theta_{23} > 0.95$$
$$\sin^2 2\theta_{13} = 0.095 \pm 0.010$$

while the current best-fit values, using direct and indirect measurements, from NuFit are:[9][10]

$$\theta_{12}[°] = 33.36^{+0.81}_{-0.78}$$
$$\theta_{23}[°] = 40.0^{+2.1}_{-1.5} \text{ or } 50.4^{+1.3}_{-1.3}$$
$$\theta_{13}[°] = 8.66^{+0.44}_{-0.46}$$
$$\delta_{CP}[°] = 300^{+66}_{-138}$$

16.1.4 Notes regarding the best fit parameter values

- These best fit values imply that there is much more neutrino mixing than there is mixing between the quark flavors in the CKM matrix (in the CKM matrix, the corresponding mixing angles are $\theta_{12} = 13.04°\pm0.05°$, $\theta_{23} = 2.38°\pm0.06°$, $\theta_{13} = 0.201°\pm0.011°$).

- These values are inconsistent with tribimaximal neutrino mixing (i.e. $\theta_{12} = \theta_{23} = 45°$, $\theta_{13} = 0°$) at a statistical significance of more than five standard deviations. Tribimaximal neutrino mixing was a common assumption in theoretical physics papers analyzing neutrino oscillation before more precise measurements were available.

- A value of θ_{23} equal to exactly 45 degrees, which would imply maximal mixing between the second and third neutrino mass eigenstates, is ruled out with a statistical significance in excess of 2 standard deviations.[10]

- The alternative choices for θ_{23} are referred to as "first quadrant" and "second quadrant" values. The data favor the first quadrant value over the second quadrant value with a statistical significance of 1.5 standard deviations in a "normal mass hierarchy" context (i.e. where the second neutrino mass eigenstate is lighter than the third neutrino mass eigenstate), but there is not a statistically significant preference between the two values in the case of an "inverted mass hierarchy" (i.e. where the second neutrino mass eigenstate is heavier than the third neutrino mass eigenstate).[10] This is the only PMNS matrix parameter which is strongly sensitive to the mass hierarchy of the neutrino masses given the currently available experimental data.[10]

- The extent to which the best fit value for δCP is meaningful should not be overstated. The best fit value for δCP is consistent with zero at the 0.9 standard deviation level, since in circular coordinates 0 degrees and 360 degrees are equivalent. Generally speaking, in particle physics, experimental results that are within 2 standard deviations of each other are called "consistent" with each other. Currently, all possible values for δCP are with 1.8 standard deviations of the best fit values, so all possible values of δCP are "consistent" with the experimental data, even though those values closer to the best fit value are somewhat more likely to be correct.

16.2 See also

- Neutrino oscillations
- Koide formula
- Cabibbo–Kobayashi–Maskawa matrix

16.3 Notes

[1] The PMNS matrix is not unitary in the seesaw model.

16.4 References

[1] Maki, Z; Nakagawa, M.; Sakata, S. (1962). "Remarks on the Unified Model of Elementary Particles". *Progress of Theoretical Physics* **28**: 870. Bibcode:1962PThPh..28..870M. doi:10.1143/PTP.28.870.

[2] Pontecorvo, B. (1957). "Inverse beta processes and nonconservation of lepton charge". *Zhurnal Éksperimental'noĭ i Teoretich-eskoĭ Fiziki* **34**: 247. reproduced and translated in *Soviet Physics JETP* **7**: 172. 1958.

[3] Kayser, Boris (February 13, 2014). "Are There Sterile Neutrinos?". arXiv:1402.3028 [hep-ph].

[4] Esmaili, Arman; Kemp, Ernesto; Peres, O. L. G.; Tabrizi, Zahra (30 Oct 2013). "Probing light sterile neutrinos in medium baseline reactor experiments". arXiv:1308.6218 [hep-ph].

[5] F.P. An, *et al.*(Daya Bay collaboration) (July 27, 2014). "Search for a Light Sterile Neutrino at Daya Bay". arXiv:1407.7259 [hep-ex].

[6] Valle, J. W. F. (2006). "Neutrino physics overview". *Journal of Physics: Conference Series* **53**: 473. arXiv:hep-ph/0608101. Bibcode:2006JPhCS..53..473V. doi:10.1088/1742-6596/53/1/031.

[7] J. Beringer *et al.* (Particle Data Group) (2012 and 2013 partial update for the 2014 edition). "PDGLive: Neutrino Mixing". Particle Data Group. Retrieved 2014-08-21. Check date values in: |date= (help)

[8]J.Beringer*et al.*(Particle Data Group) (2012). "Review of Particle Physics".*Physical Review D***86**:010001.Bibcode:a0001B. doi:10.1103/PhysRevD.86.010001.

[9] Gonzalez-Garcia, M. C.; Maltoni, M.; Salvado, J.; Schwetz, T. (June 2014). "NuFit 1.3". Retrieved 2014-07-09.

[10] Gonzalez-Garcia, M. C.; Maltoni, Michele; Salvado, Jordi; Schwetz, Thomas (21 December 2012). "Global fit to three neutrino mixing: Critical look at present precision". *Journal of High Energy Physics* **2012** (12): 123. arXiv:1209.3023. Bibcode:2012JHEP...12..123G. doi:10.1007/JHEP12(2012)123.

Chapter 17

Quark–lepton complementarity

The **quark–lepton complementarity** (**QLC**) is a possible fundamental symmetry between quarks and leptons. First proposed in 1990 by Foot and Lew,[1] it assumes that leptons as well as quarks come in three "colors". Such theory may reproduce the Standard Model at low energies, and hence quark–lepton symmetry may be realized in nature.

17.1 Possible evidence for QLC

Recent neutrino experiments confirm that the Pontecorvo–Maki–Nakagawa–Sakata matrix UPMNS contains large mixing angles. For example, atmospheric measurements of particle decay yield θPMNS
23 ≈ 45°, while solar experiments yield θPMNS
12 ≈ 34°. These results should be compared with θPMNS
13 which is small,[2] and with the quark mixing angles in the Cabibbo–Kobayashi–Maskawa matrix UCKM. The disparity that nature indicates between quark and lepton mixing angles has been viewed in terms of a "quark–lepton complementarity" which can be expressed in the relations

$$\theta_{12}^{PMNS} + \theta_{12}^{CKM} \simeq 45° \ ,$$

$$\theta_{23}^{PMNS} + \theta_{23}^{CKM} \simeq 45° \ .$$

Possible consequences of QLC have been investigated in the literature and in particular a simple correspondence between the PMNS and CKM matrices have been proposed and analyzed in terms of a correlation matrix. The correlation matrix VM is simply defined as the product of the CKM and PMNS matrices:

$$V_{\mathrm{M}} = U_{\mathrm{CKM}} \cdot U_{\mathrm{PMNS}} \ ,$$

Unitarity implies:

$$U_{\mathrm{PMNS}} = U_{\mathrm{CKM}}^{\dagger} V_{\mathrm{M}} \ .$$

17.2 Open questions

One may ask where do the large lepton mixings come from? Is this information implicit in the form of the VM matrix? This question has been widely investigated in the literature, but its answer is still open. Furthermore in some Grand

Unification Theories (GUTs) the direct QLC correlation between the CKM and the PMNS mixing matrix can be obtained. In this class of models, the V_M matrix is determined by the heavy Majorana neutrino mass matrix.

Despite the naive relations between the PMNS and CKM angles, a detailed analysis shows that the correlation matrix is phenomenologically compatible with a tribimaximal pattern, and only marginally with a bimaximal pattern. It is possible to include bimaximal forms of the correlation matrix VM in models with renormalization effects that are relevant, however, only in particular cases with $\tan\beta > 40$ and with quasi-degenerate neutrino masses.

17.3 See also

- Leptoquark

17.4 References

[1] R.Foot,H.Lew(1990). "Quark-lepton-symmetric model".*Physical Review D***41**(11):3502–3505.Bibcode:1990P.3502F. doi:10.1103/PhysRevD.41.3502.

[2] F. P. An et al. [DAYA-BAY Collaboration], Phys. Rev. Lett. 108, 171803 (2012) [arXiv:1203.1669 [hep-ex]] http://arxiv. org/abs/arXiv:1203.1669

- B.C. Chauhan, M. Picariello, J. Pulido, E. Torrente-Lujan (2007). "Quark-lepton complementarity, neutrino and standard model data predict θPMNS
  13 = (9+1
  −2)°". *European Physical Journal C* **50** (3): 573–578. arXiv:hep-ph/0605032. Bibcode:2007EPJC...50..573C. doi:10.1140/epjc/s10052-007-0212-z.

- K.M. Patel (2010). "An *SO*(10) × S_4 Model of Quark-Lepton Complementarity;". *Physics Letters B* **695**: 225. arXiv:1008.5061. Bibcode:2011PhLB..695..225P. doi:10.1016/j.physletb.2010.11.024.

17.5 Text and image sources, contributors, and licenses

17.5.1 Text

- **Flavour (particle physics)***Source:*https://en.wikipedia.org/wiki/Flavour_(particle_physics)?oldid=681888935*Contributors:*Schewek, MichaelHardy, Nurg, Xerxes314, Varlaam, Andycjp, R. fiend, DragonflySixtyseven, CALR, STGM, Andrew Gray, Knowledge Seeker, Egg, Alai,Sylvain Mielot, Linas, Mindmatrix, SpNeo, Drrngrvy, YurikBot, Bambaiah, Hairy Dude, NTBot~enwiki, Bhny, Cossy, Długosz, SCZenz,Nick, Karl Andrews, SmackBot, Incnis Mrsi, Dauto, Doug Bell, Zero sharp, Ompty, BFD1, Ruslik0, Cydebot, Hydraton31, Xxanthippe,Michael C Price, Thijs!bot, Headbomb, FelixP~enwiki, Rompe, Hayesgm, Knotwork, CosineKitty, Robin S, Askielboe, Yonidebot, Choihei,I310342~enwiki, Thecinimod, VolkovBot, A4bot, Kresadlo, Maxim, Odellus, Ptrslv72, SieBot, VVVBot, The Stickler, Muhends, PixelBot,Jtle515, Count Truthstein, DumZiBoT, MystBot, SkyLined, Addbot, ZeroOmega, SpBot, Ehrenkater, HerculeBot, Luckas-bot, Ptbotgourou,Magog the Ogre, Icalanise, Omnipaedista, Citation bot 1, Xtermin8R645, B2NVB2, Jrobbinz123, 777sms, Bizzurp, EmausBot, VinculumMan,AvocatoBot, Drift chambers, Skynden, Isambard Kingdom and Anonymous: 42
- **Quantum number** *Source:* https://en.wikipedia.org/wiki/Quantum_number?oldid=680471150 *Contributors:* The Anome, Stevertigo, Xavic69, Tim Starling, EddEdmondson, Ellywa, Mxn, Aliekens, Donarreiskoffer, Sverdrup, Hadal, Syntax~enwiki, Decumanus, Giftlite, Suspekt~enwiki, Dmmaus, Christopherlin, H Padleckas, RetiredUser2, Tsemii, Edsanville, Kareeser, Waza, Trevor MacInnis, Hidaspal, Vsmith, Spoon!, SpeedyGonsales, Nsaa, Arthena, BryanD, Pol098, Mpatel, MaximH, SeventyThree, Brownsteve, DaveTheRed, BD2412, DePiep, Happy-Camper, Fred Bradstadt, Azure8472, FlaBot, Margosbot~enwiki, Fresheneesz, Albrozdude, Bubbachuck, YurikBot, Wavelength, Mushin, Bambaiah, JWB, Hairy Dude, Huw Powell, JabberWok, Phantombantam, Jengelh, Chaos, ManoaChild, Bota47, Wknight94, Arthur Rubin, Modify, Teply, That Guy, From That Show!, Itub, SmackBot, Eskimbot, Complexica, Colonies Chris, Voyajer, Cubbi, Wybot, Yevgeny Kats, Yoshigev, Maatghandi, DJIndica, John, Sadeq, Davemcarlson, Mets501, Dan Gluck, Charles Baynham, Kushal one, Im.a.lumberjack, Imnotoneofyou, Myasuda, Cydebot, Kanags, Matrix61312, Waxigloo, SpK, Barticus88, JAnDbot, Trapezoidal, .anacondabot, Magioladitis, J.delanoy, Choihei, Acalamari, Somdebg, NewEnglandYankee, Fylwind, TraceyR, Larryisgood, AlnoktaBOT, Katoa, Go2slash, Venny85, Cobaro, EmxBot, SieBot, Gerakibot, Yintan, Allmightyduck, Agur bar Jacé, ClueBot, Keraunoscopia, Niceguyedc, Jotterbot, Versus22, Wake chaser, Joyonicity, SkyLined, Ivy martin08, Addbot, Zahd, WikiUserPedia, Jasper Deng, Drova, PV=nRT, WikiDreamer Bot, Luckas-bot, Yobot, AnomieBOT, Qmonkey, Law, Materialscientist, Citation bot, DirlBot, Melmann, Br77rino, Sahehco, A.amitkumar, FrescoBot, ProgramadorCCCP, Craig Pemberton, Redrose64, Pinethicket, Adlerbot, RobinK, Musicality213, Jordgette, Reach Out to the Truth, DARTH SIDIOUS 2, EarthCom1000, EmausBot, Gfoley4, Ravilovefriends, Demeza13, Tommy2010, Akshanshshrivastava, JSquish, ZéroBot, Mrfair, Sealbock, AManWithNoPlan, L0ngpar1sh, Just granpa, Sp4cetiger, ClueBot NG, Accelerometer, Frietjes, El.vegaro, DerekRobinson, Theopolisme, Helpful Pixie Bot, Shivsagardharam, Titodutta, Petermahlzahn, Mark Arsten, F=q(E+v^B), MrBill3, Fylbecatulous, Troller Hi, Dexbot, Webclient101, Fox2k11, Mpov, Vanamonde93, Avdhesh avistein, JaconaFrere, Septate, Tomasz59 and Anonymous: 222
- **Isospin** *Source:* https://en.wikipedia.org/wiki/Isospin?oldid=682122579 *Contributors:* Stone, Giftlite, Xerxes314, Michael Devore, RScheiber, Jason Quinn, AmarChandra, Lumidek, Perey, Rich Farmbrough, Hidaspal, V79, Cmdrjameson, RJFJR, Linas, Robert K S, Jwanders, TPickup, Ddn2, FreplySpang, Rjwilmsi, Strait, Mike Peel, Margosbot~enwiki, Goudzovski, M7bot, Bambaiah, Bhny, Archelon, Welsh, Thiseye, SmackBot, Incnis Mrsi, Sue Anne, Colonies Chris, Sawran~enwiki, KI, Iridescent, Cydebot, Michael C Price, My Flatley, Zalgo, Thijs!bot, Headbomb, Knotwork, JAnDbot, Madmarigold, Avicennasis, Lilac Soul, KIAaze, Tarotcards, Fylwind, VolkovBot, Quilbert, Anonymous Dissident, Antixt, OlekG, PaddyLeahy, SieBot, Likebox, OsamaBinLogin, Uzdzislaw, Albambot, Addbot, Luckas-bot, Citation bot, ArthurBot, Bozzochet, Obersachsebot, Glenmark, Br77rino, J04n, Ernsts, RedAcer, Citation bot 1, Minivip, FoxBot, WikitanvirBot, Helpful Pixie Bot, Bibcode Bot, BG19bot, Jamisonsloan, Monkbot, Kfitzell29, GioComitini and Anonymous: 45
- **Charm (quantum number)***Source:*https://en.wikipedia.org/wiki/Charm_(quantum_number)?oldid=669956667*Contributors:*Xavic69, Giftlite,RScheiber, Pol098, Pdelong, Bambaiah, SmackBot, Tim Q. Wells, Cydebot, Thijs!bot, Headbomb, TXiKiBoT, Count Truthstein, ArthurBot,Ulm, Ernsts, Erik9bot, Carlog3, EmausBot, CocuBot, Ibnbaja, JuhoSchultz and Anonymous: 3
- **Strangeness** *Source:* https://en.wikipedia.org/wiki/Strangeness?oldid=674819737 *Contributors:* Xavic69, Ahoerstemeier, Timwi, Herbee, Xerxes314, JeffBobFrank, RScheiber, Icairns, Xeroc, Mike Rosoft, Jkl, Jag123, LostLeviathan, Fred Condo, Mel Etitis, Tevatron~enwiki, Eyu100, Donotresus, FlaBot, Who, Fresheneesz, Srleffler, Roboto de Ajvol, Bambaiah, Hairy Dude, Conscious, Shawn81, Kyorosuke, SCZenz, 99 Willys on Wheels on the wall, 99 Willys on Wheels..., SmackBot, Stepa, JSpudeman, Complexica, Richard L. Peterson, ZICO, ShelfSkewed, Cydebot, Dchristle, Mbell, Headbomb, AntiVandalBot, NE2, The sage, I310342~enwiki, Pernogr~enwiki, Anonymous Dissident, Pamputt, Riwnodennyk, Callie.hoon, SilvonenBot, Addbot, Mr0t1633, Zorrobot, Citation bot, Wnme, Ernsts, A. di M., Qwarx, Yutsi, Johann137, Turian, Alarichus, Dinamik-bot, EmausBot, Vacation9, ???, Furkhaocean, JamesMoose, Ibnbaja and Anonymous: 33
- **Topness** *Source:* https://en.wikipedia.org/wiki/Topness?oldid=673493535 *Contributors:* Xavic69, Fredrik, RScheiber, Mako098765, Strait, Goudzovski, Bambaiah, SmackBot, Chris the speller, Tim Q. Wells, Cydebot, Headbomb, JAnDbot, Fatka, VolkovBot, Wikiwide, Count Truthstein, Addbot, Luckas-bot, ArthurBot, Ernsts, Carlog3, Craig Pemberton, DrilBot, TobeBot, Dinamik-bot, Ttsush, EmausBot, CocuBot and Anonymous: 4
- **Bottomness** *Source:* https://en.wikipedia.org/wiki/Bottomness?oldid=673493654 *Contributors:* Xavic69, Fredrik, RScheiber, Strait, Goudzovski, Bambaiah, D Monack, SmackBot, Tim Q. Wells, Cydebot, Headbomb, VolkovBot, Wikiwide, TXiKiBoT, Count Truthstein, Luckas-bot, AnomieBOT, Xqbot, Ernsts, Wetman88, Alph Bot, EmausBot, Rezabot and Anonymous: 3
- **Baryon number** *Source:* https://en.wikipedia.org/wiki/Baryon_number?oldid=675888853 *Contributors:* Andre Engels, Stevertigo, Delirium, Phys, Sanders muc, Securiger, Herbee, Xerxes314, Dratman, RScheiber, Jason Quinn, Discospinster, Pjacobi, Brim, Guy Harris, H2g2bob, Linas, Ted BJ, Isnow, Ddn2, BD2412, Raymond Hill, Bubba73, Margosbot~enwiki, Fresheneesz, Cannywizard, PointedEars, Roboto de Ajvol, YurikBot, Bambaiah, Tom Lougheed, V1adis1av, QFT, Doug Bell, Dan Gluck, Cydebot, Thijs!bot, Headbomb, Barakitty, Richard n, JAnDbot, CosineKitty, Bbi5291, Siryendor, FaTTshady74, STBotD, TXiKiBoT, A4bot, Venny85, SieBot, Muhends, Erodium, Addbot, WikiDreamer Bot, Legobot, Luckas-bot, Amirobot, ArthurBot, XZeroBot, Ernsts, MastiBot, Ttsush, Sahimrobot, Ernest3.141 and Anonymous: 17
- **Lepton number***Source:*https://en.wikipedia.org/wiki/Lepton_number?oldid=678945693*Contributors:*Xavic69, Alfio, Phys, Robbot, RScheiber,Jag123, Flying fish, SeventyThree, Yurik, Bambaiah, JabberWok, That Guy, From That Show!, SmackBot, Incnis Mrsi, Fuhghettaboutit, Doug

Bell, Khazar, Man pl, Blinking Spirit, Cydebot, Thijs!bot, Headbomb, WolfmanSF, Leyo, FaTTshady74, A4bot, Murkee, Gerakibot, Peachypoh, Addbot, Peti610botH, Yobot, Amirobot, AnomieBOT, Unara, Citation bot, GrouchoBot, 🈁🈁🈁, HRoestBot, EmausBot, Timetraveler3.14, ChuispastonBot, Bibcode Bot, BG19bot, Monkbot, MarkovianStumble and Anonymous: 24

- **Weak isospin** *Source:* https://en.wikipedia.org/wiki/Weak_isospin?oldid=679239808 *Contributors:* Xavic69, Charles Matthews, Giftlite, RScheiber, Hidaspal, Ian Pitchford, Chobot, Roboto de Ajvol, YurikBot, Bambaiah, Jimp, RussBot, Paul D. Anderson, Jheriko, Bbabba, Cydebot, Michael C Price, Headbomb, Igodard, Tokei-so, Andre.holzner, Pamputt, AlleborgoBot, Muhends, L.smithfield, Tvine, Addbot, Icalanise, ArthurBot, Ernsts, Puzl bustr, John of Reading and Anonymous: 17

- **Electric charge** *Source:* https://en.wikipedia.org/wiki/Electric_charge?oldid=681550518 *Contributors:* AxelBoldt, Mav, Andre Engels, Roadrunner, Peterlin~enwiki, Heron, JohnOwens, Michael Hardy, Ixfd64, Delirium, Looxix~enwiki, Ellywa, Mdebets, Glenn, Rossami, Nikai, Andres, Raven in Orbit, Reddi, Omegatron, Gakrivas, Lumos3, Rogper~enwiki, Gentgeen, Robbot, Fredrik, Dukeofomnium, Wikibot, Fuelbottle, Wjbeaty, Giftlite, DavidCary, Herbee, Snowdog, Dratman, Valen~enwiki, RScheiber, Jason Quinn, Brockert, OldakQuill, Manuel Anastácio, LiDaobing, Karol Langner, Icairns, Iantresman, GNU, Vincom2, Discospinster, Guanabot, Jpk, Dbachmann, ZeroOne, Laurascudder, Bobo192, Rbj, Giraffedata, Kjkolb, Scentoni, Mdd, Alansohn, Atlant, ABCD, Velella, Wtshymanski, HenkvD, Mikeo, DV8 2XL, Gene Nygaard, HenryLi, Oleg Alexandrov, Nuno Tavares, Cimex, Rocastelo, StradivariusTV, Oliphaunt, BillC, Eleassar777, Cyberman, Palica, BD2412, Demonuk, Edison, SMC, Krash, Dougluce, FlaBot, Psyphen, Nivix, Alfred Centauri, Gurch, Kri, Gdrbot, Manscher, YurikBot, Bambaiah, Lucinos~enwiki, Stephenb, Manop, Pseudomonas, JDoorjam, TDogg310, Chichui, Kkmurray, Wknight94, Light current, Enormousdude, Johndburger, Tcsetattr, Pinikas, Reyk, Canley, Geoffrey.landis, JDspeeder1, GrinBot~enwiki, Mejor Los Indios, Sbyrnes321, Marquez~enwiki, Moeron, Vald, Thunderboltz, Dmitry sychov, HalfShadow, Gilliam, Oscarthecat, Andy M. Wang, Chris the speller, Lenko, DHN-bot~enwiki, Dual Freq, Hallenrm, Rrburke, The tooth, MichaelBillington, Hgilbert, Drphilharmonic, Daniel.Cardenas, Springnuts, Yevgeny Kats, Andrei Stroe, DJIndica, Naui~enwiki, Nmnogueira, SashatoBot, Richard L. Peterson, Slowmover, Cronholm144, Mgiganteus1, Bjankuloski06en~enwiki, Nonsuch, Ben Moore, RandomCritic, MarkSutton, Stikonas, Dicklyon, Levineps, Igoldste, Tawkerbot2, Chetvorno, JForget, CmdrObot, Kehrli, Jsd, Myasuda, Cydebot, Fl, Bvcrist, Meno25, Gogo Dodo, WISo, Christian75, Ssilvers, Thijs!bot, Epbr123, Barticus88, N5iln, Mojo Hand, Headbomb, Gerry Ashton, Escarbot, Aadal, AntiVandalBot, Seaphoto, Prolog, DarkAudit, Lyricmac, Tim Shuba, WikifingHelper, Asgrrr, JAnDbot, Acroterion, Bongwarrior, VoABot II, J2thawiki, Sstolper, Jjurik, Bubba hotep, User A1, DerHexer, InvertRect, Robin S, MartinBot, M. Bilal Shafiq, LedgendGamer, Pharaoh of the Wizards, Numbo3, Hans Dunkelberg, Night-Falcon90909, Uncle Dick, Ginsengbomb, Katalaveno, DarkFalls, NewEnglandYankee, QuickClown, Juliancolton, ACBest, Treisijs, Lseixas, Jefferson Anderson, Sheliak, Philip Trueman, TXiKiBoT, The Original Wildbear, Ayan2289, Nickipedia 008, LuizBalloti, Monty845, Jpalpant, Biscuittin, Demmy100, SieBot, Gerakibot, Caltas, Gastin, Wing gundam, Msadaghd, JerrySteal, Jojalozzo, Oxymoron83, Faradayplank, Avnjay, Anchor Link Bot, Neo., Loren.wilton, ClueBot, The Thing That Should Not Be, Arakunem, Termine, Mild Bill Hiccup, Stephaninator, LeoFrank, Excirial, Kocher2006, Jusdafax, Brews ohare, Cenarium, Jotterbot, PhySusie, SchreiberBike, Wuzur, JDPhD, Versus22, Thinking Stone, Rror, Cernms, Truthnlove, Addbot, Some jerk on the Internet, CanadianLinuxUser, NjardarBot, LaaknorBot, Scottyferguson, LinkFA-Bot, Naidevinci, Ocwaldron, Tide rolls, Lightbot, JDSperling, Legobot, Luckas-bot, Yobot, CinchBug, Duping Man, AnomieBOT, DemocraticLuntz, Sertion, Jim1138, IRP, Pyrrhus16, Kingpin13, Bluerasberry, Materialscientist, Geek1337~enwiki, ImperatorExercitus, Xqbot, TheAMmollusc, Phazvmk, Addihockey10, Capricorn42, Nnivi, ProtectionTaggingBot, RibotBOT, Srr712, A. di M., Constructive editor, Frozenevolution, Ryryrules100, Jc3s5h, Drunauthorized, Mithrandir, Steve Quinn, Bevelstbug, Fast kartwheels, BenzolBot, DivineAlpha, AstaBOTh15, Pinethicket, I dream of horses, Jivee Blau, Calmer Waters, Tinton5, MastiBot, Serols, Meaghan, Lalrang2007, Logical Gentleman, FoxBot, TobeBot, ScheryP, Jonkerz, Ndkartik, Vrenator, Taytaylisious09, Ammodramus, Jamietw, DARTH SIDIOUS 2, Eshmate, Irfanyousufdar, EmausBot, John of Reading, GoingBatty, K6ka, Darkfight, Hhhippo, JSquish, Harddk, Stephen C Wells, Liam McM, Sonygal, L Kensington, Donner60, Peter Karlsen, Sven Manguard, Planetscared, ClueBot NG, Jack Greenmaven, Cking1414, Ihwood, Ulflund, CocuBot, MelbourneStar, O.Koslowski, Brickmack, AvocatoBot, Rm1271, Altaïr, F=q(E+v^B), Snow Blizzard, Brad7777, Bhaskarandpm, Eduardofeld, GoShow, Dexbot, JoshyyP, Brandonsmacgregor, Reeceyboii, Frosty, Reatlas, I am One of Many, Eyesnore, Tentinator, Germeten, Nablacdy, Spyglasses, Freddyboi69, 20M030810, SpecialPiggy, Marizperoj, Peterfreed, Rigid hexagon, Jiteshkumar727464, Dyeith, Podayeruma, Oleaster, Layfi, BlueDecker, GeneralizationsAreBad, Pritam kumar Barik, KasparBot, Ramprakashsfc and Anonymous: 427

- **X (charge)** *Source:* https://en.wikipedia.org/wiki/X_(charge)?oldid=605281425 *Contributors:* XJaM, BD2412, SmackBot, Michael C Price, Headbomb, Addbot, Xqbot, Ernsts, A. di M. and Anonymous: 1

- **Hypercharge** *Source:* https://en.wikipedia.org/wiki/Hypercharge?oldid=681928054 *Contributors:* Xavic69, Schneelocke, Charles Matthews, Phys, Robbot, Filemon, Giftlite, JeffBobFrank, Mike Rosoft, Pjacobi, Jag123, Xayma, Jörg Knappen~enwiki, Xauw, Roboto de Ajvol, YurikBot, Wavelength, Bambaiah, Hairy Dude, Gaius Cornelius, Paul D. Anderson, KnightRider~enwiki, SmackBot, Tom Lougheed, Dauto, V1adis1av, Radagast83, CRGreathouse, CmdrObot, Michael C Price, Headbomb, Zylorian, Alphachimpbot, JAnDbot, Robin S, HEL, Jnnnnn, STBotD, Anonymous Dissident, Venny85, Antixt, COBot, Zenohm, SkyLined, Addbot, Luckas-bot, Amirobot, Götz, Citation bot, Svix, Wiles stunlalt, MastiBot, DixonDBot, EmausBot, Preon, Halfb1t, ChrisGualtieri, Lesser Cartographies, SJ Defender and Anonymous: 20

- **Weak hypercharge** *Source:* https://en.wikipedia.org/wiki/Weak_hypercharge?oldid=679239006 *Contributors:* Xavic69, Pjacobi, MeltBanana, David Schaich, BD2412, Chobot, Roboto de Ajvol, YurikBot, Bambaiah, Paul D. Anderson, Incnis Mrsi, Dauto, Michael C Price, Headbomb, Andre.holzner, HEL, Anonymous Dissident, Pamputt, Antixt, Tvine, MystBot, Addbot, Luckas-bot, Yobot, Götz, Ernsts, Slightsmile, QuantumSquirrel, ResidentAnthropologist, Helpful Pixie Bot, Bambi12~enwiki, EzP z4 and Anonymous: 20

- **Cabibbo–Kobayashi–Maskawa matrix***Source:*https://en.wikipedia.org/wiki/Cabibbo%E2%80%93Kobayashi%E2%80%93Maskawa_matrix?oldid=663541514*Contributors:*Michael Hardy, Charles Matthews, DJ Clayworth, BenRG, MathMartin, Giftlite, Xerxes314, Rich Farm-brough, Hidaspal, RJHall, Army1987, Cmdrjameson, Jag123, Physicistjedi, Mennato, RJFJR, AndyBuckley, Oleg Alexandrov, Linas, Ruzik-lan, Rjwilmsi, Mathbot, Itinerant1, Goudzovski, YurikBot, Bambaiah, Jimp, JabberWok, Pseudomonas, Black Falcon, RG2, Teply, Grin-Bot~enwiki, SmackBot, Jim62sch, Stepa, Hmains, Dauto, QFT, Ligulembot, Lambiam, Sipos0, CmdrObot, Markus Pössel, ...ﻉ ﻟﺍﻲﺍ ﻥﻮﺟ ﻢﻴﻛ ﻥﺎﺑﻭﺎﻟﻞﻟﺍ!ﺍ, Headbomb, I310342~enwiki, Drgnrave, Barraki, Cuzkatzimhut, VolkovBot, Anonymous Dissident, Pamputt, J-ishikawa~enwiki,Jim E. Black, Quasirandom, PipepBot, NuclearWarfare, DumZiBoT, TimothyRias, Addbot, Mjamja, Lightbot, PV=nRT, Luckas-bot, Yobot,AnomieBOT, Citation bot, TechBot, A. di M., FrescoBot, Citation bot 1, E-Soter, Bphyswiki, Brandmeister, Ebehn, Isocliff, PhysicsAboveAll,Widr, Fiddleyhead, Bibcode Bot, Nikos Papadakis, Sushant.2811, Monkbot and Anonymous: 39

- **Pontecorvo–Maki–Nakagawa–Sakata matrix***Source:*https://en.wikipedia.org/wiki/Pontecorvo%E2%80%93Maki%E2%80%93 Nakagawa%E2%80%93Sakata_matrix?oldid=677906442*Contributors:*Julesd, Jordi Burguet Castell, Giftlite, Bender235, Strait, Goudzovski, Ohwilleke,

Leo C Stein, Hydraton31, Headbomb, R'n'B, Cuzkatzimhut, TXiKiBoT, FourteenDays, Jasondet, Copyeditor42, Addbot, Debresser, Luckas-bot, Yobot, Citation bot, Blennow, Ace111, A. di M., D'ohBot, Puzl bustr, John of Reading, ZéroBot, ClueBot NG, Bibcode Bot, Franzl aus tirol, Foreveriii, Danholly and Anonymous: 15

- **Quark–lepton complementarity** *Source:* https://en.wikipedia.org/wiki/Quark%E2%80%93lepton_complementarity?oldid=667466538 *Contributors:* Chuunen Baka, RJFJR, Stephenb, SmackBot, BryanG, JorisvS, Cydebot, BetacommandBot, Koeplinger, Headbomb, Mentifisto, Ya-hel Guhan, STBotT, Ergo leu, Paulfharrison, BartekChom, TubularWorld, Yobot, Omnipaedista, Citation bot 1, Bibcode Bot, BattyBot and Anonymous: 8

17.5.2 Images

- **File:Ambox_current_red.svg** *Source:* https://upload.wikimedia.org/wikipedia/commons/9/98/Ambox_current_red.svg *License:* CC0 *Contributors:* self-made, inspired by Gnome globe current event.svg, using Information icon3.svg and Earth clip art.svg *Original artist:* Vipersnake151, penubag, Tkgd2007 (clock)

- **File:Ambox_important.svg** *Source:* https://upload.wikimedia.org/wikipedia/commons/b/b4/Ambox_important.svg *License:* Public domain *Contributors:* Own work, based off of Image:Ambox scales.svg *Original artist:* Dsmurat (talk · contribs)

- **File:Baryon-decuplet-small.svg** *Source:* https://upload.wikimedia.org/wikipedia/commons/7/78/Baryon-decuplet-small.svg *License:* Public domain *Contributors:* Own work *Original artist:* Trassiorf

- **File:Baryon-octet-small.svg** *Source:* https://upload.wikimedia.org/wikipedia/commons/b/b5/Baryon-octet-small.svg *License:* Public domain *Contributors:* Own work *Original artist:* Trassiorf

- **File:Baryon_decuplet_w_mass.png** *Source:* https://upload.wikimedia.org/wikipedia/en/c/c1/Baryon_decuplet_w_mass.png *License:* PD *Contributors:*
self-made
Original artist:
Venny85 (talk)

- **File:Baryon_octet_w_mass.png** *Source:* https://upload.wikimedia.org/wikipedia/en/9/98/Baryon_octet_w_mass.png *License:* PD *Contributors:*
self-made
Original artist:
Venny85 (talk)

- **File:Bcoulomb.png** *Source:* https://upload.wikimedia.org/wikipedia/commons/0/04/Bcoulomb.png *License:* Public domain *Contributors:* http://en.wikipedia.org/wiki/Image:Bcoulomb.png *Original artist:* ?

- **File:Cabibbo_angle.svg** *Source:* https://upload.wikimedia.org/wikipedia/commons/5/50/Cabibbo_angle.svg *License:* Public domain *Contributors:* Own work *Original artist:* Headbomb

- **File:Copyright-problem.svg** *Source:* https://upload.wikimedia.org/wikipedia/en/c/cf/Copyright-problem.svg *License:* PD *Contributors:* ?
Original artist: ?

- **File:Edit-clear.svg** *Source:* https://upload.wikimedia.org/wikipedia/en/f/f2/Edit-clear.svg *License:* Public domain *Contributors:* The *Tango! Desktop Project*. *Original artist:*
The people from the Tango! project. And according to the meta-data in the file, specifically: "Andreas Nilsson, and Jakub Steiner (although minimally)."

- **File:Eg1.png** *Source:* https://upload.wikimedia.org/wikipedia/en/b/be/Eg1.png *License:* PD *Contributors:*
self-made
Original artist:
Venny85 (talk)

- **File:Eg2.png** *Source:* https://upload.wikimedia.org/wikipedia/en/3/3e/Eg2.png *License:* PD *Contributors:*
self-made
Original artist:
Venny85 (talk)

- **File:Eg3.png** *Source:* https://upload.wikimedia.org/wikipedia/en/1/15/Eg3.png *License:* PD *Contributors:*
self-made
Original artist:
Venny85 (talk)

- **File:Eg4.png** *Source:* https://upload.wikimedia.org/wikipedia/en/6/65/Eg4.png *License:* PD *Contributors:*
self-made
Original artist:
Venny85 (talk)

- **File:Electric_field_point_lines_equipotentials.svg** *Source:* https://upload.wikimedia.org/wikipedia/commons/9/96/Electric_field_point_lines_equipotentials.svg *License:* Public domain *Contributors:* Own work *Original artist:* Sjlegg

17.5.3 Content license

www.ingramcontent.com/pod-product-compliance
Lightning Source LLC
Chambersburg PA
CBHW070942180526
45168CB00003B/1148